Statistical Rules of Thumb

Statistical Rules of Thumb

Gerald van Belle

WILEY-INTERSCIENCE

A JOHN WILEY & SONS, INC. PUBLICATION

Library of Congress Cataloging-in-Publication Data:

van Belle, Gerald.
    Statistical rules of thumb / Gerald van Belle.
        p. cm. — (Wiley series in probability and statistics)
    Includes bibliographical references and index.
    ISBN 0-471-40227-3 (pbk : alk. paper)
    1. Mathematical statistics.   I. Title.   II. Series
QA276.12 .V36 2001
519.5—dc21   2001026954

Printed in the United States of America

10 9 8

To Johanna

"Our science is like a store filled with the most subtle intellectual devices for solving the most complex problems, and yet we are almost incapable of applying the elementary principles of rational thought. In every sphere, we seem to have lost the very elements of intelligence: the ideas of limit, measure, degree, proportion, relation, comparison, contingency, interdependence, interrelation of means and ends."

Simone Weil, from "The Power of Words," in *Simone Weil: An Anthology* by Sian Miles (editor), Grove Press, New York, page 222 (1986).

Contents

Preface   *xiii*

Acronyms   *xvii*

1 *The Basics*   *1*

1.1   Distinguish Randomized and Observational Studies   2

1.2   Beware of Linear Models   3

1.3   Understand Omnibus Quantities   6

1.4   Independence, Equal Variance, and Normality   7

1.5   Models As Simple As Possible, But Not More Simple   11

1.6   Do Not Multiply Probabilities More Than Necessary   12

1.7   Know the Sample Space for Statements of Risk   13

1.8   Use Two-sided p-Values   14

1.9 p-Values for Sample Size, Confidence Intervals for
      Results   16

1.10 Use at Least Twelve Observations in Constructing a
      Confidence Interval   18

1.11 Know the Unit of the Variable   19

1.12 Know Properties Preserved When Transforming Units   20

1.13 Be Flexible About Scale of Measurement Determining Analysis — 23

1.14 Be Eclectic and Ecumenical in Inference — 24

1.15 Consider Bootstrapping for Complex Relationships — 25

1.16 Standard Error from Sample Range/Sample Size — 26

2 Sample Size — 29

2.1 Begin with a Basic Formula for Sample Size — 31

2.2 No Finite Population Correction for Survey Sample Size — 33

2.3 Calculating Sample Size Using the Coefficient of Variation — 35

2.4 Do Not Formulate a Study Solely in Terms of Effect Size — 38

2.5 Overlapping Confidence Intervals Do Not Imply Nonsignificance — 39

2.6 Sample Size Calculation for the Poisson Distribution — 40

2.7 Sample Size for Poisson With Background Rate — 41

2.8 Sample Size Calculation for the Binomial Distribution — 43

2.9 When Unequal Sample Sizes Matter; When They Don't — 45

2.10 Sample Size With Different Costs for the Two Samples — 47

2.11 The Rule of Threes for 95% Upper Bounds When There Are No Events — 49

2.12 Sample Size Calculations Are Determined by the Analysis — 50

3 Covariation — 53

3.1 Assessing and Describing Covariation — 55

3.2 Don't Summarize Regression Sampling Schemes with Correlation — 56

3.3 Do Not Correlate Rates or Ratios Indiscriminately — 58

3.4 Determining Sample Size to Estimate a Correlation — 59

3.5 Pairing Data is not Always Good — 61

3.6 Go Beyond Correlation in Drawing Conclusions — 63

3.7	Agreement As Accuracy, Scale Differential, and Precision	65
3.8	Assess Test Reliability by Means of Agreement	68
3.9	Range of the Predictor Variable and Regression	70
3.10	Measuring Change: Width More Important than Numbers	72

4 Epidemiology **75**

4.1	Start with the Poisson to Model Incidence or Prevalence	76
4.2	The Odds Ratio Approximates the Relative Risk Assuming the Disease is Rare	77
4.3	The Number of Events is Crucial in Estimating Sample Sizes	82
4.4	Using a Logarithmic Formulation to Calculate Sample Size	84
4.5	Take No More than Four or Five Controls per Case	86
4.6	Obtain at Least Ten Subjects for Every Variable Investigated	87
4.7	Begin with the Exponential Distribution to Model Time to Event	89
4.8	Begin with Two Exponentials for Comparing Survival Times	91
4.9	Be Wary of Surrogates	92
4.10	Prevalence Dominates in Screening Rare Diseases	95
4.11	Do Not Dichotomize Unless Absolutely Necessary	99
4.12	Select an Additive or Multiplicative Model on the Basis of Mechanism of Action	100

5 Environmental Studies **103**

5.1	Think Lognormal	103
5.2	Begin with the Lognormal Distribution in Environmental Studies	104
5.3	Differences Are More Symmetrical	106
5.4	Beware of Pseudoreplication	108

5.5	Think Beyond Simple Random Sampling	109
5.6	Consider the Size of the Population Affected by Small Effects	111
5.7	Statistical Models of Small Effects Are Very Sensitive to Assumptions	112
5.8	Distinguish Between Variability and Uncertainty	113
5.9	Description of the Database is As Important as Its Data	115
5.10	Always Assess the Statistical Basis for an Environmental Standard	116
5.11	Measurement of a Standard and Policy	117
5.12	Parametric Analyses Make Maximum Use of the Data	119
5.13	Distinguish Between Confidence, Prediction, and Tolerance Intervals	120
5.14	Statistics Plays a Key Role in Risk Assessment, Less in Risk Management	122
5.15	Exposure Assessment is the Weak Link in Assessing Health Effects of Pollutants	124
5.16	Assess the Errors in Calibration Due to Inverse Regression	125

6 Design, Conduct, and Analysis   **129**

6.1	Randomization Puts Systematic Effects into the Error Term	129
6.2	Blocking is the Key to Reducing Variability	131
6.3	Factorial Designs Should be Used to Assess Joint Effects of Variables	132
6.4	High–Order Interactions Occur Rarely	134
6.5	Balanced Designs Allow Easy Assessment of Joint Effects	136
6.6	Analysis Follows Design	137
6.7	Plan to Graph the Results of an Analysis	139
6.8	Distinguish Between Design Structure and Treatment Structure	142
6.9	Make Hierarchical Analyses the Default Analysis	143

6.10 Distinguish Between Nested and Crossed Designs—
      Not Always Easy   145

6.11 Plan for Missing Data   146

6.12 Address Multiple Comparisons Before Starting the
      Study   149

7 Words, Tables, and Graphs   **153**

7.1 Use Text for a Few Numbers, Tables for Many
      Numbers, Graphs for Complex Relationships   153

7.2 Arrange Information in a Table to Drive Home the
      Message   155

7.3 Always Graph the Data   158

7.4 Never Use a Pie Chart   160

7.5 Bargraphs Waste Ink; They Don't Illuminate Complex
      Relationships   162

7.6 Stacked Bargraphs Are Worse Than Bargraphs   163

7.7 Three-Dimensional Bargraphs Constitute Misdirected
      Artistry   166

7.8 Identify Cross-sectional and Longitudinal Patterns in
      Longitudinal Data   167

7.9 Use Rendering, Manipulation, and Linking in High
      Dimensional Data   170

8 Consulting   **175**

8.1 Structure a Consultation Session to Have a Beginning,
      a Middle, and an End   176

8.2 Ask Questions   177

8.3 Make Distinctions   178

8.4 Know Yourself, Know the Investigator   180

8.5 Tailor Advice to the Level of the Investigator   181

8.6 Use Units the Investigator is Comfortable With 182

8.7 Agree on Assignment of Responsibilities   184

8.8 Any Basic Statistical Computing Package Will Do 185

8.9 Ethics Precedes, Guides, and Follows Consultation 186

8.10 Be Proactive in Statistical Consulting   187

8.11 Use the Web for Reference, Resource, and Education   189

8.12 Listen to, and Heed the Advice of Experts in the Field   190

Epilogue   **193**

References   **195**

Author Index   **207**

Topic Index   **211**

Preface

This book presents statistical rules of thumb for practitioners of statistical science. The intent is stimulate thinking about designing and conducting a study, analyzing its data, or communicating its results. The book should be accessible to those who have had a basic course in statistics. It should be useful to researchers, graduate students in one of the quantitative sciences, and scientists who want to understand the statistical underpinnings of research efforts.

The book begins with basic statistical considerations about inference, assumptions and statistical processes. The next two chapters deal with sample size and covariation. Chapters 4 and 5 deal with particular areas: epidemiology and environmental studies. The choice of these topics is based on my lifelong association with these areas. Chapter 6 deals with the design of studies and Chapter 7 with the presentation of data. These two chapters are self-contained; start with them to get quick ideas about design, analysis and presentation of research results. Chapter 8 completes the book by dealing with statistical consulting. In a sense, this brings us back to the topic of Chapter 1. The chapters can be read independently of each other with the exception of Chapter 2 which provides a basis for much of the remainder of the book. Most of the equations in the book are numbered, though they are not always referenced. This was done in order to facilitate discussion of the rules on the web site.

One of the goals I set for the book was to provide a justification for every rule. This could be based on the implications of a statistical model, or practical experience. Every rule is discussed under five headings: introduction, statement of the rule, illustration, basis of the rule, and discussion and extensions.

A book begins as a gleam in an author's eye and, like starlight, may take years to reach the viewer. This book is no exception. It goes back at least 10 years to a conversation with Bea Shube, then an editor at Wiley. In a letter to me from 1992 she said, "I want it, I want it, I want it!" Since then I have collected statistical rules of thumb in a nonsystematic and nonrandom fashion. It's a little late, but here it is.

It was the enthusiastic encouragement of Steve Millard that started a serious attempt at writing. He also suggested the format for each rule and produced a preliminary draft of Chapter 1. Tom Louis, a statistician's statistician, made many suggestions and shrewd observations. Jim Hughes has been a colleague for more than 20 years providing suggestions and comments on many aspects of the book. Mathematics graduate student Shawn Cokus helped me with the intricacies of LATEX—giving the stuff the manuals don't tell: how to make it work on your machine. He made it look so easy that I had to ask him twice in some cases. Lance Jolley generated the initial graphs for Chapter 2; he did proofreading and provided overall support.

The National Research Center for Statistics and the Environment (Peter Guttorp, director) provided partial support for a year. As a visitor to NRCSE during 1998-1999 I worked on this book (among other things). Working with Peter Guttorp, Paul Sampson, and Mary Lou Thompson was always, and continues to be, a pleasure.

Ed Gehan provided the suggestion and reference to the rule about estimating the standard error from the range of the observations. Scott Emerson convinced me about the virtues of a power of 0.975. Clarice Weinberg shared a preprint of an article on p-values. I suspect that in spite of her wise remarks it won't be the end of the controversy.

Several classes of students in statistical consulting were given parts of the book. It's always exciting to share with students and get their creative and unjaded comments.

Roger Higdon and Bud Kukull read Chapter 4 making insightful comments. Margaret Pepe was extremely helpful with comments on screening, and other comments on Chapter 4. Jackie Benedetti and Nancy Temkin read and commented on Chapter 8. Michael Levitz helped me during the most crucial part of the writing: when graphs just would not come out right. He did most of the graphs and worked on the indexing. One of my sons, Louis, provided one of the graphs in Chapter 7 and also helped me set up the Web site.

The editors at Wiley, Steve Quigley, Heather Haselkorn, and Lisa Van Horn were helpful and straightforward. Amy Hendrickson provided excellent LATEX advice and produced the final camera ready copy. It's been a pleasure to work with them.

My colleagues in the departments of biostatistics, environmental health, statistics, and epidemiology provide a concentration of expertise and commitment that is phenomenal. It is at national meetings that you realize what an extraordinary group of faculty have been assembled at the University of Washington. I particularly want to single out Lloyd Fisher. Although he did not comment on the book directly he has been my life-long collaborator and friend. He represents for me the ideal combination of scholarship and integrity.

A person who has been my thought-companion all my professional life is Daniel B. DeLury (now deceased). I took my first statistics courses from him at the University of Toronto. His aphoristic statements still ring in my ears: "A frequency distribution is a quality of permanence not inherent in the items themselves," and "Randomization puts systematic effects into the error term." Already in 1963 he was worried about "statistical analyses untouched by the human mind," surely even more of a worry forty years later.

Paul Crane read the manuscript after it was in "final" form. His suggestions for improvement were so perceptive and valuable that I spent several weeks implementing them. There are few pages that do not bear his constructive and critical markings.

Having acknowledged tremendous assistance and encouragement, it should be clear that I am responsible for the final content of the book. Mistakes are mine, not the friends listed above.

I expect that many readers will look for their favorite rule of thumb and not find it. Please accept my apologies. You may quarrel with a particular rule and feel you have a better rule. In both cases, send me an e-mail and I will incorporate it on my Web site (http://www.vanbelle.org). I plan to post a Rule of Thumb of the Month on this site. Perhaps a second volume can be written; a deep lode of statistical wisdom can be mined, and it is my hope that this book will stir a collective desire to bring this wisdom to the surface.

This book would not have seen the light of day without the loving, quiet but firm, encouragement of my wife, Johanna. This book is dedicated to her.

GERALD VAN BELLE

January 15, 2002
Seattle, Washington

Acronyms

Acronym	Definition, page
AIDS	Acquired immunodeficiency syndrome, 25
AIRS	Aerometric information retrieval system, 115
ANOVA	Analysis of variance, 15
AR(1)	First order autoregressive process, 8
ARE	Asymptotic relative efficiency, 99
ASA	American Statistical Association, 186
CAST	Cardiac Arrhythmia Suppression Trial, 93
C	Cost of a study, 48
CD4	Cluster of differentiation (on T4 cells), 25
CV	Coefficient of variation, 35
DLR	Diagnostic likelihood ratio, 98
EPA	Environmental Protection Agency, 103
EPV	Events per variable, 88
FDA	Food and Drug Administration, 2
FEV_1	Forced expiratory volume after 1 second, 107
GEE	Generalized estimating equations, 22

GLMM	Generalized linear mixed models, 22	
HEI	Health Effects Institute, 104	
HIV	Human immunodeficiency virus, 92	
IMS	Institute of Mathematical Statistics, 175	
IRT	Item response theory, 70	
ISI	International Statistical Institute, 186	
MEGO	My eyes glaze over, 182	
NMMAPS	National Morbidity and Mortality Air Pollution Study, 111	
NGO	Non-governmental organization, 104	
NRC	National Research Council, 104	
NUATRC	(Mickey Leland) National Urban Air Toxics Research Center, 104	
PBPK	Physiologically based pharmacokinetic, 5	
PC	Proportionate change, 35	
ppm	parts per million, 139	
PCA	Principal component analysis, 88	
PCB	Polychlorinated biphenyl, 14	
PM	Particulate matter, 111	
PREV	Prevalence, 95	
PVC	Premature ventricular contraction, 93	
NPV	Predictive value negative test, 96	
PPV	Predictive value positive test, 96	
ROC	Receiver-operator characteristic, 98	
SAS	Statistical analysis system (original name), 136	
SE	Standard error, 26	
SIDS	Sudden infant death syndrome, 78	
SENS	Sensitivity, 96	
SPEC	Specificity, 96	
STAR	Science to achieve results, 103	
TSP	Total suspended particulates, 118	
URL	Universal (Uniform) resource locator, 189	
WEB	Short version of "World Wide Web", 189	

Statistical Rules of Thumb

1
The Basics

This chapter discusses some fundamental statistical issues dealing with variation, statistical models, calculations of probability and the connection between hypothesis testing and estimation. These are basic topics that need to be understood by statistical consultants and those who use statistical methods. The selection of these topics reflects the author's experience and practice.

There would be no need for statistical methods if there were no variation or variety. Variety is more than the spice of life; it is the bread and butter of statisticians and their expertise. Assessing, describing and sorting variation is a key statistical activity. But not all variation is the domain of statistical practice, it is restricted to variation that has an element of randomness to it.

Definitions of the field of statistics abound. See a sampling in Fisher and van Belle (1993). For purposes of this book the following characteristics, based on a description by R.A. Fisher (1935) will be used. Statistics is the study of populations, variation, and methods of data reduction. He points out that "the same types of problems arise in every case." For example, a population implies variation and a population cannot be wholly ascertained so descriptions of the population depend on sampling. The samples need to be reduced to summarize information about the population and this is a problem in data reduction. It will be very helpful to see how any statistical problem involves the definition by asking:

- What is the population that has to be described?

- What are the sources of variation?

- How should the data be reduced?

The use of randomness in some form enables the statistician to use probability theory. So there is a need for understanding some of this theory in order to carry out statistical analysis. This book will make considerable use of elementary probability theory to validate statistical rules of thumb.

1.1 DISTINGUISH BETWEEN RANDOMIZED AND OBSERVATIONAL STUDIES

Introduction

One of the basic distinctions to make in designing or analyzing studies is between randomized and observational studies. Lack of randomness leads to "arm waving": asserting that an inference is correct, in spite of the lack of randomness.

Rule of Thumb

Make a sharp distinction between observational and experimental studies.

Illustration

In survey sampling a "convenience sample" is often used. By this is meant a nonrandom sample. A standard criticism of such a sample is that it is not "representative" meaning that statements about the sample cannot be applied to the population of interest. The standard reply is "Yes, but. . .." This then leads to an attempted justification of the validity of making the inference. At times the attempt will be valid, at times it will not. The listener winds up making a judgment about the validity of the argument—not the sample.

Basis of the Rule

The Achilles' heel of statistical practice and inference is the selection of the sample. A probability model is required to draw valid inferences.

Discussion and Extensions

The phrases "observational data" and "observational studies" imply an unknown probabilistic pedigree. Since the data are considered a sample, the interest is really in the population represented by the sample. In these situations, precision is less of an issue than bias. Concato et al. (2000) and Benson and Hartz (2000) argue that carefully carried out observational studies compare favorably with randomized clinical trials—and that such studies may be the only ones ethically possible. Epidemiologists are experts at conducting careful observational studies. However, regulatory agencies such as the United States Food and Drug Administration (FDA) will continue to insist on evidence from randomized clinical trials in the drug evaluation process. The analysis of "nonexperimental" data has fascinated statisticians for as long as statistics has been considered a scientific discipline. A good summary of recent approaches can be found in Copas and Li (1997) and the discussion of this paper.

The argument has been made that for nonrandom samples from a population a "descriptive" analysis rather than an "inferential" analysis should be made. This misses the point, it is almost certainly the case (an inference) that the sample itself is not of interest but the population from which it came. Hence, inferences will be made; the challenge to the statistician is to guide this process by pointing out appropriate cautions. The practice of medicine provides an excellent example of inference executed on nonrandom samples. It also indicates that it is a combination of art and science—as is the practice of statistics.

Another terminology talks about a "representative sample." This is an unusually slippery phrase. Several eminent statisticians have tried to get their arms around it— taking four substantial papers of steadily increasing length to do so (see Kruskal and Mosteller, 1979a, 1979b, 1979c, and 1980.

While the majority of studies are—and will continue to be— observational; on balance, the gold standard continues to be studies that incorporate randomization in some form. Randomization is important not only for a balancing of underlying contributing factors but also for a valid basis for statistical procedures.

1.2 BEWARE OF LINEAR MODELS

Introduction

Statisticians are trained to fit statistical models to data; they are particularly good at fitting linear models and generalizations. It is salutary to note the following observation by March (1996) "... finding an empirical formula that fits the data but is not based on any underlying theory does little to advance our understanding."

Rule of Thumb

Always look for a physical model underlying the data being analyzed. Assume that a statistical model such as a linear model is a good first start only.

Illustration

Suppose that Galileo had generated pairwise data consisting of the time it took for a cannon ball to fall a certain distance. He carried out this experiment by dropping the ball from various heights of the tower of Pisa so that T (time in seconds) was the outcome variable and S (distance in meters) the predictor variable. A least squares equation is fitted to the data. This is a reasonable approach, and the main source of variability will be in the measurement of T since S can be measured without appreciable error.

Table 1.1 Height from which cannonball was dropped and time to reach the ground

Height (meters)	Time (seconds)	Height (meters)	Time (seconds)
10	1.8	35	2.7
10	1.2	35	3.0
15	1.6	40	2.7
15	1.7	40	2.5
20	2.2	45	2.9
20	2.2	45	2.8
25	2.2	50	2.6
25	2.2	50	3.0
30	2.7	55	3.3
30	2.2	55	3.7

Suppose the data looked like the observations listed in Table 1.1. A statistician analyzes the data with a "first approximation" linear model. The least squares line produces:

$$\text{Time (seconds)} = 1.243 + 0.0375 \text{ Height (meters)},$$

with both intercept and slope highly significant ($p \leq 0.001$). A plot of residuals shows good agreement with the linear model. If the analyst stops here, and many do, the underlying structure of the relationship will have been missed. Further, more extensive data would have shown that the assumed linear relationship does not reflect the underlying physical reality of the gravitational law.

Basis of the Rule

It is crucial to have a broad understanding of the subject matter involved. Statistical analysis is much more than just carrying out routine computations. Only with keen understanding of the subject matter can statisticians, and statistics, be most usefully engaged.

Discussion and Extensions

Models can be viewed and used at three levels (Ford, 2000). The first is a model that fits the data. A test of goodness-of-fit operates at this level. This level is the least useful but is frequently the one at which statisticians and researchers stop. For example, a test of a linear model is judged good when a quadratic term is not significant. A second level of usefulness is that the model predicts future observations. Such a model has been called a *forecast model* (Guttorp, 1995). This level is often required in screening studies or studies predicting outcomes such as growth rate.

A third level is that a model reveals unexpected features of the situation being described, a *structural model* (Guttorp, 1995). Going back to the illustration, the linear model certainly fits the data and can predict new values with reasonable accuracy— within a limited range. However, it does not explain the data. To achieve that, the model must be inverted to come up with the well-known physical equation:

$$S = \frac{1}{2}gT^2. \tag{1.1}$$

Practitioners rarely deal with models at this level even though it has probably the widest scientific implications. Models describing birth and death processes tend to be of this type. Another example, from pharmacology, would be physiologically based pharmacokinetic (PBPK) models. A survey of statistical practice would probably show that the majority of models used are at the first level.

The example illustrates many aspects of measurement and statistical issues. First of all, the physical model suggests that distance is a function of time but the experiment fixes distance and looks at time. This suggests fitting the model:

$$T = \sqrt{\frac{2}{g}S}, \tag{1.2}$$

a linear model without intercept. The estimate and confidence interval for the slope can then be translated into an estimate of g and associated confidence interval (producing $\hat{g} = 10.2$ and 95% confidence interval 9.8–11.3, which straddles the known value of 9.81 m/sec^2). This calculation also illustrates that confidence intervals constructed in one scale and then, assuming a monotonic transformation, can be expressed in a new scale. A final point, the two models considered by equations (1.1) and (1.2) are non-hierarchical and it is difficult to compare them, especially since the scales of the models differ.

Wilkinson (1999) gives sound advice:

> ...some researchers who use statistical methods pay more attention to goodness of fit than to the meaning of a model. It is not always convenient to remember that the right model for a population can fit a sample of data *worse* than a wrong model—even a wrong model with fewer parameters. Statisticians cannot rely on statistical diagnostics to save them, especially with small samples. They must think about what the models mean,regardless of fit, or they will promulgate nonsense.

R.L. Chambers (1997), a discussant of the Copas and Li paper (Copas and Li, 1997), points out that statisticians should identify and integrate two modeling activities: modeling of the population and second, modeling of the data selection process from the population. The first step is frequently assumed or known ("...let X be a random normal deviate..."). In observational studies, identification of the first step is critical for valid inference. Copas and Li (1997) also provide a nice example of how to integrate selection bias explicitly into a model. For additional discussion, see Rules 1.5 and 5.8.

The key to good statistical practice is understanding the subject matter to which the methodology is being applied. This knowledge enables the statistician to recommend and evaluate appropriate classes of models and possible alternatives.

1.3 UNDERSTAND OMNIBUS QUANTITIES

Introduction

Data reduction is an important task for statisticians. There are two kinds of data reduction. The first type involves calculations of statistics that estimate parameters, such as the sample mean to estimate the population mean. The second type of data reduction is based on model reduction. For example, the standardized difference between two normal populations with the same variance,

$$\Delta = \frac{\mu_1 - \mu_2}{\sigma}, \tag{1.3}$$

reduces the number of parameters from three to one. Other examples are p-values, Mahalanobis distances, and F-tests for analyses of variance. Such parameters and statistics can be called *omnibus quantities*.

Rule of Thumb

Be sure to understand the components and purpose of an omnibus quantity.

Illustration

A very popular omnibus quantity is *effect size*. The quantity Δ above is a parametric example. It combines the difference in population means with a measure of population variation. An analogous quantity can be constructed from statistics. While this quantity has intuitive appeal and wide applicability it should be not absolutized as the key goal of a research project.

Basis of the Rule

Omnibus quantities represent many-one mappings. Thus for the same value of this quantity many component values are possible. Ordinarily the components are of primary scientific interest, while the omnibus quantity is a shorthand summary of intermediary interest.

Discussion and Extensions

The virtue of omnibus quantities is that they provide one-number summaries. This is also their weakness. Beside the many-one mapping problem, there is the additional problem that estimates of such quantities are dependent on the design of the study— for example, consider comparison of two populations based on paired data versus a comparison based upon two independent samples. Both designs provide a valid estimate of the population difference. However, how will the variability be estimated?

To insist on a particular formulation of the effect size may rule out otherwise desirable designs. A rigid, one-size-fits-all prescriptive commitment to a particular omnibus quantity (such as effect size) puts researchers in unnecessary and unprofitable straight-jackets.

Another good example of omnibus quantification involves data in 2×2 contingency tables. The number of ways of looking at these simple four numbers is astonishing. One-number summaries include the odds ratio, the relative risk, the kappa statistic, the chi-square statistic, Youden's J statistic, Yule's Q statistic, and many others. Two number summaries include sensitivity, specificity; positive and negative predictive values of tests; proportions in column and row margins. An example of a three–number summary is sensitivity, specificity, prevalence. Assuming that the data are proportions that sum to 1, this three-number summary is strictly speaking not a summary but a re-expression. Selection of a specific summary statistic will depend on the way the data were generated and the purpose of the summary.

Omnibus quantities are omnipresent. They fulfil a useful, summarizing role. They are less useful in understanding basic mechanisms. Often they are constructed to achieve invariance under rescalings or relocations. The researcher should determine before-hand that such transformations do not compromise the objectives of the analysis—for example, in measuring agreement, statistics that are invariant under rescaling and relocation are not appropriate (see Rule 3.7 on page (65) for further discussion).

1.4 ASSESS INDEPENDENCE, EQUAL VARIANCE, AND NORMALITY—IN THAT ORDER

Introduction

Classical hypothesis tests involve making assumptions about the underlying distributions from which the data were sampled. Usually, the only difference in these assumptions between parametric and nonparametric tests is the assumption of a normal distribution; all the other assumptions for the parametric test apply to the nonparametric counterpart as well.

Rule of Thumb

Classical hypothesis tests assume that the observations (1) are independent, (2) all come from populations with the same variance, and, for parametric tests, (3) follow a normal distribution. The most important (in terms of maintaining an assumed Type I error level: the probability of rejecting the null hypothesis when it is true—see introduction to Chapter 2 for a fuller discussion of errors) are the first, then the second, then the third.

Illustration

If observations are taken serially—for example, sequential laboratory measurements—the investigator should be first of all concerned about serial correlation, then homogeneity of variance, then normality. Frequently, a histogram of the observations will be made, "to make sure the observations are normal," but no attempt is made to assess correlation among the observations.

Basis of the Rule

Many investigators have studied the issue of the relative importance of the assumptions underlying hypothesis tests (see, Cochran, 1947; Gastwirth and Rubin, 1971; Glass et al., 1972; Lissitz and Chardos, 1975; Millard et al., 1985; Pettitt and Siskind, 1981; Praetz, 1981; Scheffé, 1959; and others). All studied the effects of correlated errors on classical parametric or nonparametric tests. In all of these studies, positive correlation resulted in an inflated Type I error level, and negative correlation resulted in a deflated Type I error level. The effects of correlation were more important than differences in variances between groups, and differences in variances were more important than the assumption of a normal distribution.

Discussion and Extensions

Consider the one-sample t-test of the null hypothesis vs. the one-sided alternative hypothesis. The test of this hypothesis is based on the t-statistic:

$$t = \frac{\bar{x}}{s/\sqrt{n}} \tag{1.4}$$

When the observations are independent, the standard error of $\bar{x}$ is $\sigma/\sqrt{n}$ and is estimated by the denominator of the t-statistic. Suppose now that the observations can be ordered (say in time) and are correlated as follows: Adjacent observations have correlation ρ, observations one step apart have correlation ρ^2, and so on. This is called a first order autoregressive process, AR(1), with correlation ρ. In this case it can be shown that for reasonably large n,

$$\text{True standard error of } \bar{x} \approx \sqrt{\frac{1+\rho}{1-\rho}} \frac{s}{\sqrt{n}} \tag{1.5}$$

(see Millard et al., 1985; and Lettenmaier,1976). The true standard error is larger than the calculated standard error when $\rho > 0$ and smaller when $\rho < 0$. Thus when $\rho > 0$, use of unadjusted $s/\sqrt{n}$ leads to a t-statistic that is too large and that increases chance of a Type I error. The effect is substantial even for relatively small ρ.

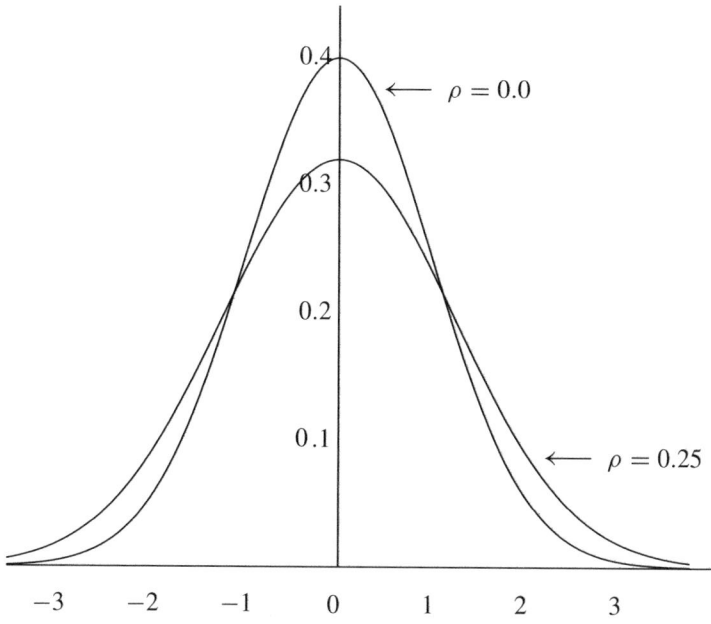

Fig. 1.1   Distribution of t-statistic for Uncorrelated and Positively Correlated Data.

Figure 1.1 illustrates the wider spread of autocorrelated data with autocorrelation $\rho = +0.25$ demonstrating the increased Type I error associated with positive correlation. For reasonably large n,

$$t \sim N\left(0, \sqrt{\frac{1+\rho}{1-\rho}}\right). \tag{1.6}$$

The effect of this kind of autocorrelation on the Type I error can be seen in Table 1.2.

Table 1.2   Effect of Autocorrelation on Type I Error—Assumed 0.05 for $\rho = 0$

ρ	$\sqrt{\frac{1+\rho}{1-\rho}}$	Adjusted Value of t-Statistic	Type I Error
0	1	1.96	0.05
0.1	1.11	1.77	0.08
0.2	1.22	1.61	0.11
0.3	1.36	1.44	0.15
0.4	1.53	1.28	0.20
0.5	1.73	1.13	0.26

Table 1.2 shows that even modest autocorrelation increases the chance of a Type I error substantially. The adjusted value is the t-statistic (1.96) divided by $\sqrt{\frac{1+\rho}{1-\rho}}$. For an autocorrelation of $\rho = 0.2$ the probability of a Type I error is more than doubled. Of course, with an increased Type I error there is a decreased Type II error and, hence, increased power. While true, this makes a shambles of the usual hypothesis testing framework. Inspection of Table 1.2 shows that for modest correlation (say, $\rho < 0.25$) there needs to be a $100\rho\%$ reduction in the t-statistic to get the correct estimate of the Type I error. A similar table can be constructed for negative autocorrelation, which will result in a smaller Type I error. The quantity $\sqrt{\dfrac{1+\rho}{1-\rho}}$ is also the amount the usual standard error of the mean has to be multiplied by in order to get the true standard error. For a correlation of $\rho = 0.2$ the standard error has to be inflated by 22%. These calculations demonstrate the importance of considering autocorrelation in statistical inference.

A similar discussion can be found in Scheffé (1959), who focuses on a simpler autocorrelated process–but with comparable results.

David (1985), in what must be the shortest paper ever, shows that

$$0 \le E(s^2) \le \frac{n}{n-1}\sigma^2, \qquad (1.7)$$

where s^2 is the sample variance of n observations, not necessarily independent, E is the expected value, and σ^2 is the population variance. If there is dependence, there can be substantial underestimation of the population variance. The discussion in this section leads to the further conclusion that randomization in experimental design is very important because by randomization some kinds of serial correlation can be eliminated.

The second condition for the validity of tests of hypotheses is that of homogeneity of variance. Box (1953) already showed that hypothesis tests are reasonably robust against heterogeneity of variance. For a two-sample test a three-fold difference in variances does not affect the probability of a Type I error. Tests of equality of variances are very sensitive to departures from the assumptions and usually don't provide a good basis for proceeding with hypothesis testing. Box (1953) observed that,

> ...to make the preliminary test on variances is rather like putting to sea in a rowing boat to find out whether conditions are sufficiently calm for an ocean liner to leave port!

Normality is the least important in tests of hypotheses. It should be noted that the assumption of normality deals with the error term of the model, not the original data. This is frequently forgotten by researchers who plot histograms of the raw data rather than the residuals from the model. Graphical tests for normality, such as normal probability plots, involving small sample sizes, are difficult to interpret. Such small samples can generate what would seem to be large deviations from normality. There are simple tests based on correlation of the ranked observations with normal order statistics. An accessible reference is Looney and Gulledge (1985), see also some correspondence generated by this paper for additional references.

Independence of observations is clearly the most important assumption of a test of a hypothesis. It is also the most difficult one to evaluate. This is one more example where knowledge of the subject area is crucial, and a thorough understanding of the way the data were generated can lead to a valid statistical analysis.

1.5 KEEP MODELS AS SIMPLE AS POSSIBLE, BUT NOT MORE SIMPLE

Introduction

The selection or choice of a statistical model is an important part of data analysis and data reduction (see Rule 1.2, for an example). By what criteria will such data be reduced? There is a famous rule called Ockham's razor, which provides a guide for the selection of models. The rule states that, everything else being equal, the simpler model among a collection of models is to be preferred.

Rule of Thumb

One formal statement of Ockham's razor is: "Do not multiply hypotheses beyond necessity."

Illustration

Ockham's razor in statistical analysis is used implicitly when models are embedded in richer models—for example, when testing the adequacy of a linear model by incorporating a quadratic term. If the coefficient of the quadratic term is not significant, it is dropped and the linear model is assumed to summarize the data adequately.

Basis of the Rule

There is no formal statistical basis for this rule; its validity and applicability is a meta-scientific construct based on the intuitively appealing idea that "simpler is better." The word *parsimony* incorporates the basic thrust of Ockham's razor. The rule is the basis of Einstein's statement that explanations of scientific phenomena should be as simple as possible, but not more simple.

Discussion and Extensions

There are several formulations of Ockham's razor. Adams (1987) lists four:

- "It is futile to do with more what can be done with fewer."

- "When a proposition comes out true for things, if two things suffice for its truth, it is superfluous to assume a third."

- "Plurality should not be assumed without necessity."

- "No plurality should be assumed unless it can be proved (a) by reason, or (b) by experience, or (c) by some infallible authority."

Ockham's razor has been used in very broad contexts—for example, reductionism. Any non-juridical statement beginning with "just," hinges on a form of Ockham's razor, as in the following: "A brake is just a device for transforming kinetic energy into heat." No doubt, a brake transforms kinetic energy into heat, but there are a host of other contexts and qualifications of a brake.

The theory of hierarchical models makes use of Ockham's razor: simpler models are embedded in a hierarchical fashion in more complicated ones. The situation becomes more intricate when competing models are not hierarchical, such as comparing an exponential with a polynomial model. Ockham's razor cannot be applied immediately to this situation. Another razor can be constructed, such as comparing the models by the logarithm of the likelihoods.

There is a tension between Ockham's razor and Rule 1.2 which warns about the use of linear models. The tension can be creative, forcing the researcher to contemplate the underlying mechanism of the phenomena being investigated. The decision, then, whether or not use a linear model will be less data-driven and more dependent on the investigator's insights and creativity.

Ockham's razor involves the concept of parsimony, which is another way of expressing economy of scientific concepts and models. All these terms indicate that there is nothing static about data, analysis, and models. There is dynamic interplay among these three. Statistics provides a methodology for conducting research.

1.6 DO NOT MULTIPLY PROBABILITIES MORE THAN NECESSARY

Introduction

Probabilities are sometimes multiplied indiscriminately. This section proposes a razor for dealing with probabilities.

Rule of Thumb

Do not multiply probabilities beyond necessity.

Illustration

This example deals with a 1964 purse snatching in Los Angeles, California as described in Paulos (1988). A "blond woman with a ponytail snatched a purse from another woman. The thief fled on foot but was later spotted entering a yellow car driven by a black man with a beard and a mustache." A woman was finally located who matched this description. The prosecution assigned probabilities to the above facts: blond hair (1/3), ponytail (1/10), yellow car (1/10), black man with beard (1/10), moustache (1/4), and interracial couple in car (1/1000). These probabilities were then multiplied together to produce a probability of 1/12,000,000. This probability was calculated under the null hypothesis of no association and was judged so low that the court convicted the couple.

The probability could have been made even smaller by creatively adding additional characteristics. Note also, that multiplication of these probabilities assumes that the events were independent.

Basis of the Rule

Probabilities are bounded by 1; multiplication of enough probabilities will always lead to a small number.

Discussion and Extensions

It should be noted that the conviction was appealed to the California Supreme Court and was overturned (see Paulos, 1988, pages 39–42, for further discussion). Beside the dubious practice of introducing more and more events into the calculations, multiplication of probabilities implies that the underlying events are assumed to be independent. Both of these conditions need more than just casual attention. This cautionary note about the multiplication of probabilities is particularly important when no action (or action) is based on a small probability. For example, the probability of failure of, say, a bridge can be made arbitrarily small by adding more and more "independent" events.

Anyone multiplying probabilities needs to address the nature of probability. For example, "the probability of life on Mars" derived from multiplying probabilities associated with conditions necessary for life needs to have the sample space explicated. This is where views of probability will determine whether such probabilities, and their multiplication is warranted or not. See Rule 1.14 for additional discussion.

To summarize, there are two problems with the indiscriminate multiplication of probabilities. First, multiplication without adjustment implies that the events represented by the probabilities are treated as independent. Second, since probabilities are always less than 1, the product will become smaller and smaller. If small probabilities are associated with unlikely events then, by a suitable selection, the joint occurrence of events can be made arbitrarily small.

1.7   KNOW THE SAMPLE SPACE FOR STATEMENTS OF RISK

Introduction

A statement of probability requires, at least, a sample space over which the statement is made. One of the first things learned about a probability is that it is conditional.

Rule of Thumb

Make sure the sample space is known for a statement of risk.

Example

The probability of dying in a mountain climbing accident is estimated to be about 0.0006 according to Crouch and Wilson (1982). This is the annual risk, presumably in the group of people who climb mountains. However, if a person never climbs a mountain the risk will be 0.

Basis of the Rule

A probability is associated with a sample space.

Discussion and Extensions

A probability can frequently be expressed as a ratio of the number of events divided by the number of units eligible for the event. What the rule of thumb says is to be aware of what the numerator and denominator are, particularly when assessing probabilities in a personal situation. If someone never goes hang gliding, they clearly do not need to worry about the probability of dying in a hang gliding accident.

It is important to understand why the probability statement is being made. In the court case example, Rule 1.6, the probability was being applied to the defendants; that is, they were assumed to be part of the denominator and the numerator. From the defense's point of view, it was important to know where the denominator and numerator came from. In other words, what was the relevant subset? The actual choice of the subset is not easy to establish.

Another example: A state official is doing a risk assessment of eating shell fish caught in a local bay known to have polychlorinated biphenyl (PCB) contamination. The risk assessment applies only to people who are likely to eat shell fish and likely to eat locally caught shell fish. The risk is zero for a person who does neither one. A general population-based calculation may be made if the public cost of the potential illness from PCB is desired. These examples indicate that it is important to specify what probabilities are being calculated and for what purpose.

Every probability involves a sample space. Creators of probabilities and their intended audiences need to be crystal clear about the underlying universe of discourse. An automatic, reflex question, "What group, units, tests etc. is this probability dealing with?" will save a great deal of time and frustration.

1.8   USE TWO-SIDED *P*-VALUES

Introduction

The use of one-sided or two-sided *p*-values is a subject of some controversy. It basically deals with the issue of whether the alternative hypothesis is one-sided or two-sided. A two-sided hypothesis usually involves a two-sided rejection region.

Rule of Thumb

The use of one-sided p-values is discouraged. Ordinarily, use two-sided p-values.

Illustration

In a clinical trial the researchers assume that the active treatment is always equal or better than the control treatment. Hence their p-values are calculated on the basis of a one-sided test of significance. Empirical evidence frequently shows such one-sided alternatives to be wrong.

Basis of the Rule

The concept of equipoise in hypothesis testing usually implies that both sides of the null hypothesis are considered equally plausible. If the null hypothesis is false, it is not known which direction the relationship will go.

Discussion and Extensions

The terminology in this section is loose. A p-value is associated with a test of significance rather than a test of hypothesis. However, most statisticians use the concepts interchangeably: a one-sided alternative hypothesis is associated with the calculation a one-sided p-value.See the introduction to Chapter 2 for a fuller delineation of the hypothesis testing framework.

The statistical and epidemiological literature contains extensive discussions about the use of one-sided and two-sided p-values. The discussion centers on the possible rejection region of the null hypothesis, that is, the status of the alternative hypothesis. For example, in comparing a new drug with a placebo, the null hypothesis usually is that there is no effect. The alternative hypothesis is that the drug has an effect, and therefore the response should be in the positive direction. It is argued that the new drug must be at least as good as the placebo, and the rejection region for the alternative hypothesis should be one-sided. This assumption is debated on empirical grounds. Particularly in the area of clinical trials, one-sided hypothesis tests are rare.

It is not uncommon to have researchers argue for a one-sided test because the two-sided observed p-value happens to be between 0.05 and 0.10. Hence, if the magical 0.05 limit is to be reached the one-sided test will achieve this. This practice borders on the unethical; an ex post facto argument for a one-sided p-value is always suspect. If the researcher is confronted with the logical extension of the one-sided test that even if the results are "significant" in the other direction cannot be claimed, the insistence on a one-sided test is frequently dropped.

There are two situations where a one-sided test is logical. First, in statistical genetics, tests of genetic linkage are naturally modeled as the linkage parameter being greater than zero (only). Second, it is not always appreciated that the F-test in the analysis of variance (ANOVA) is one-sided. The reason for its validity is that the underlying alternative hypothesis is still two-sided.

If a researcher argues for a one-sided test as a matter of principle, then it is recommended to set the probability of the Type I error at, say, 0.025. This will maintain a two-sided test of 0.05 and yet satisfy the researcher's principled concerns. This was done in the case of the Cardiac Antisuppressant Trial because it was assumed that an approved drug, used for further indications, could only benefit the patient. In fact, the drug was shown to be harmful. See the discussion in Friedman et al. (1998) and Rule 4.9.

Ordinarily, it is best to avoid one-sided alternative hypotheses and associated p-value procedures. One-sided tests also lead to one-sided confidence intervals (see the next section) and they are much less accepted.

1.9 USE P-VALUES TO DETERMINE SAMPLE SIZE, CONFIDENCE INTERVALS TO REPORT RESULTS

Introduction

Certain journals present tables of p-values (or, even worse, F-statistics, degrees of freedom and associated p-values). Others downplay the use of p-values and insist on presentation of statistical estimates and associated confidence intervals. What is a good rule of thumb? The discussion can be framed in terms of hypothesis testing and estimation.

Rule of Thumb

When designing experiments or observational studies focus on p-values to calculate sample size; when reporting results focus on confidence intervals.

Illustration

Consider Tables 1.3 and 1.4, which are part of the results of studies; the first table expresses the results in terms of hypothesis testing, the second in terms of estimation. It is clear that Table 1.4 is much more informative.

Basis of the Rule

Hypothesis testing and estimation are two sides of the same coin. Hypothesis testing is somewhat more formal in that the alternative hypothesis must be defined if a sample size calculation is to be carried out. The data ingredients in these two approaches are the same. One way to think about this is that in hypothesis testing the investigator "stands" on the null hypothesis and see where the statistic is located. In estimation, the investigator is located on the statistic and sees where the hypothesis value is located (several hypotheses could even be contemplated!). In both cases the p-value is equivalent to the distance between the hypothesis and the statistic.

Table 1.3 Pairwise Comparison of Treatments: Hypothesis Testing Approach

Comparison	Statistic	p-Value
Treatment 1 vs. control	$F(1,20) = 6.53$	$p < 0.05$
Treatment 2 vs. control	$F(1,20) = 5.21$	$p < 0.05$
Treatment 1 vs. Treatment 2	$F(1,20) = 1.29$	$p > 0.05$

Table 1.4 Pairwise Comparison of Treatments: Estimation Approach

Comparison	Difference	95% Confidence Interval
Treatment 1 vs. control	1.25	(0.90, 1.60)
Treatment 2 vs. control	1.40	(1.05, 1.75)
Treatment 1 vs. Treatment 2	−0.15	(−0.50, 0.20)

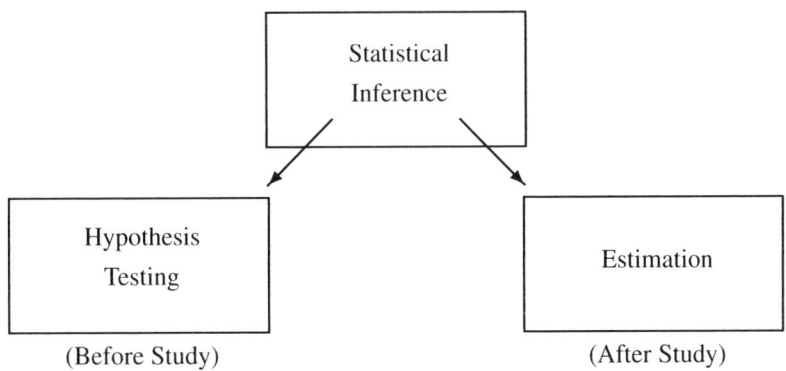

Discussion and Extensions

If the hypothesis testing information is considered crucial it would be possible to use the asterisk convention: "*" means $0.01 < p < 0.05$, "**" means $0.005 < p < 0.01$, and "***" means $0.001 < p < 0.005$. Values that are not significant can be denoted "n.s." Other schemes are possible; just let the reader know what the convention is. Note also that the third line of Table 1.4 indicates that the third estimate is derivable from the first two, hence does not add new information but only reformulates the first two comparisons.

The criticism of the use of p-values can go too far. There is a legitimate use for p-values. For example, in stepwise regression procedures the p-values provide a formal, public framework for the entry and deletion of variables. Particularly in routine testing situations it is useful to have such a framework. There are logical problems associated with stepwise regression, Yandell (1997) suggests using this procedure as a basis for beginning model exploration, not ending it.

A careful discussion of p-values can be found in Oakes (1990). A very negative view can be found in Rothman (1986). A later reflection can be found in Weinberg (2001).

There is universal agreement about the need to calculate sample sizes for proposed research in order to know the cost and effort required. This requires some sort of hypothesis testing framework. It is also agreed that results need to be summarized by more than just p-values. Thus, hypothesis testing and confidence interval approaches are complementary.

1.10 USE AT LEAST TWELVE OBSERVATIONS IN CONSTRUCTING A CONFIDENCE INTERVAL

Introduction

Precision does not vary linearly with increasing sample size. As is well known, the width of a confidence interval is a function of the square root of the number of observations. But it is more complicated than that. The basic elements determining a confidence interval are the sample size, an estimate of variability, and a pivotal quantity associated with the estimate of variability. Specifically, for samples from a normal distribution the sample standard deviation has an associated pivotal variable, the $t-$ statistic. (A *pivotal variable* is an omnibus quantity consisting of parameters and statistics, and has a distribution that does not depend on the parameters.)

Rule of Thumb

The width of a confidence interval, involving estimation of variability and sample size, decreases rapidly until 12 observations are reached and then decreases less rapidly.

Illustration

Figure 1.2 provides the estimated half-width of a confidence interval for a population mean assuming the sample came from a normal population. Specifically, the graph is a plot of $\dfrac{t_{n-1,1-\alpha/2}}{\sqrt{n}}$ against the sample size. Up to a sample size of about 12 there is a rapid drop.

Basis of the Rule

The reciprocal of the square root provides a non-linear drop-off. A second contributor is the the decreasing value of the t-statistic for fixed probabilities, as the sample size increases.

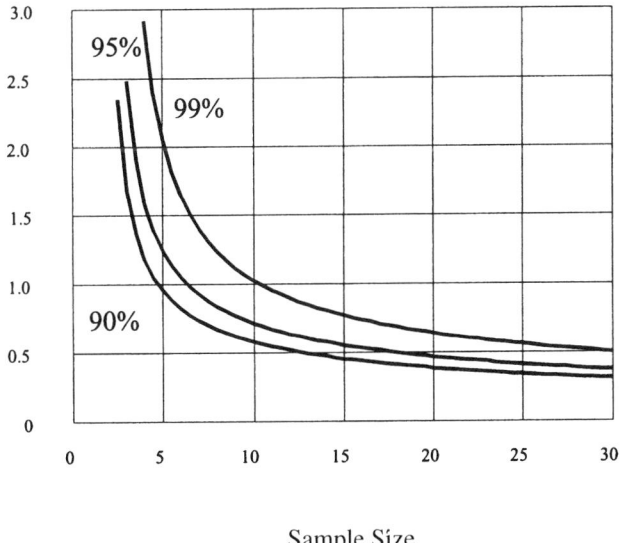

Sample Size

Fig. 1.2 Half-width confidence interval assuming a t-statistic with $n - 1$ degrees of freedom and sample size n. Confidence levels of 90%, 95%, and 99.

Discussion and Extensions

Figure 1.2 can also be used to calculate the half-width confidence interval associated with any standard deviation. The ordinate is the number of standard deviations needed to arrive at a specified half-width and confidence level as indicated on the figure. For example, for a sample size of $n = 15$ the multiplier for a 95% confidence interval is about 0.5. If the standard deviation associated with a sample of 15 observations is 30, then the half-width of a 95% confidence interval is $0.5 \times 30 = 15$. This is a good rule to remember (the actual value is 0.56). At $n = 30$ the multiplier for a 95% half-width confidence interval is 0.37, representing a rather slow decrease from the value at $n = 15$.

Figure 1.2 indicates that the efficiency of sampling effort can depend on more than just the sample size for small samples. If the variance has to be estimated from the data, the uncertainty due to this estimation, which is incorporated in the t-statistic, decreases also as the sample size increases.

1.11 KNOW THE UNIT OF THE VARIABLE

Introduction

Variables come with units. Sometimes these units are considered "natural." For example, annual income is usually measured in dollars or euros or pounds. It is not "natural" to consider the logarithm of this income. However, statisticians frequently transform variables in order to achieve homogeneity of variance, normality, or some other statistical property.

Rule of Thumb

Always know what the unit of a variable is.

Illustration

There are urban legends that deal with assumptions of units. One such story is that a Canadian commercial airline pilot misinterpreted the amount of jet fuel on board by assuming that the quantity was expressed in gallons when, in fact, it was liters. Another example is the loss of a Mars probe due to a confusion of metric and non-metric standards. There is a whole science of units that deals with making certain that units are carried through correctly in complicated physical derivations.

Basis of the Rule

Defining units explicitly is a scientific courtesy that fosters sound communication.

Discussion and Extensions

"Natural" units are not always simple. For example, the pH scale and the Richter scale are fairly complicated constructs. By dint of repetition, they have become familiar and therefore "natural." For measuring temperature there have been many scales developed. The three most common ones are Centigrade, Fahrenheit, and Kelvin. This fact illustrates that different units may be used for different purposes. Units are frequently assumed and misunderstood.

1.12   KNOW PROPERTIES PRESERVED WHEN TRANSFORMING UNITS

Introduction

Transformations of units has a long history in statistics. A basic distinction is between linear and non-linear transformations. Transforming units in degrees Centigrade to degrees Fahrenheit is an example of a linear transformation ($^0F = {}^0C(\frac{9}{5}) + 32$). Examples of nonlinear transformations are the square root, logarithmic and, logistic transformations. In theoretical statistics there are other purposes of transformations such as achieving statistical independence, but these will not be discussed. See Kendall et al. (1987) for a discussion of these transformations.

A great deal of statistical modeling involves partitioning of observed variability of a response into two parts,

$$\text{Response} = \text{Fixed or Systematic Effects} + \text{Random Effects.} \qquad (1.8)$$

This is similar to the engineering concept of the response being a mixture of signal and noise, and the aim to decompose the two. A fundamental aim of statistical analysis is to sort the response into the fixed and random components with the ideal that the random component does not depend on the fixed component (homogeneity of variance), has nice distributional properties (for example, normality), is simply added to the fixed component (additivity), and does not exhibit systematic effects (statistical independence). Transformations have the aim of achieving some, or all, of the above. They usually involve a compromise, achieving these goals only partially.

In line with the above, immediate aims of transformations are to (1) stabilize the variance of a statistic, (2) make the distribution of a variable more symmetric, and (3) make the distribution of the statistic, or the error term, more normal. In the first part of the twentieth century these were very important considerations in order to reduce computations. This is less of an issue these days.

A fourth aim involves linkage of the probability of a response (often a binary event such as failure and success) to an underlying model. For example, the logistic model discussed in Chapter 4. The transformation is an example of a probability integral transformation. One effect of this transformation is to transform the interval [0,1] into the real line $[-\infty, +\infty]$.

Rule of Thumb

Know what properties a transformation preserves and does not preserve.

Illustration

The normal distribution has the property that a linear transformation of a normal random variable is normal. That is, if Y is normally distributed, then $Y - 1$ is also normally distributed. This is not true for a linear transformation of a Poisson random variable. In fact, the sample space is completely changed. For example, if the random variable Y is Poisson, then $Y - 1$ will not be Poisson. The sample space now includes the value -1, which was not the case for the untransformed variable.

Basis of the Rule

Transformations are not innocuous. Like some medications, if they work they also have side effects. Read the "fine print." Like other medications, there may be side effects, and they may not work.

Discussion and Extensions

The two most common uses for transformations are to achieve homogeneity of variance and/or to come closer to normality. In the best cases both are achieved with the same transformation. For example, the square root of a Poisson variable stabilizes the variance and makes the distribution more symmetrical. Transformations differ in the extent to which they modify a variable or a parameter. The square root transformation deforms to a lesser extent than the logarithmic transformation. The usual rule is that a square root transformation is appropriate for variables with variance proportional to the mean (as in the Poisson case), and a logarithmic transformation is needed when the standard deviation is proportional to the mean (as in the lognormal distribution). The determination of the appropriate transformation is often determined by a plot of either the variances or standard deviations of samples against their means. Whichever graph is more linear then determines the appropriate transformation. A formal approach can be applied using the Box and Cox transformation; see, for example, Kendall et al. (1987) or any text on statistics. An example of such a transformation is equation (4.33), page 101.

Some transformations do not preserve ordering of summary statistics. Consider the following two data sets: {1,10} and {5,5}. The arithmetic means are 5.5 and 5.0, respectively. The geometric means (equivalent to a logarithmic transformation) are 3.2 and 5.0, respectively. Hence the transformation did not preserve the rank ordering of the arithmetic means.

As will be indicated in the rule dealing with bootstrapping, Rule 1.15, there is much less need for transformations today. That is, to accomplish the inferential goal there are other techniques that are now available.

Nelder and Wedderburn (1972) introduced the concept of *link*. Originally, the fixed part of the response was assumed to be linear in the parameters. Nelder and Wedderburn generalized the systematic part by allowing a large class of functions to constitute the systematic part. For example,

$$\text{Response} = e^{\alpha + \beta X} + \text{Random.} \tag{1.9}$$

In this this model the exponential link is intrinsically nonlinear, that is, it is not possible to transform the model to get one linear in the parameters *and* retain the additive structure of the random component (taking logarithms of the response destroys the additivity of the random component). These models and their extensions are amazingly flexible and have eliminated the need for transformations in many cases. They have led to Generalized Estimating Equation (GEE) models by Liang and Zeger (1986) and Generalized Linear Mixed Models (GLMM) by Breslow and Clayton (1993). An excellent discussion of transformations can be found in Atkinson and Cox (1988).

A check list of the effects of a transformation should include the following:

- What is the purpose of the transformation?

- What is the effect of the transformation on the sample space of the transformed variable or parameter?

- Were the original responses represented by additive link of the systematic and the random components of the model? If so, a nonlinear transformation will destroy this relationship.

- Did the original response have a normal error associated with it? If so, then a nonlinear transformation will destroy normality.

There is still a place for the kinds of transformations discussed in this section. In the following chapters many transformations will be used to simplify models, stabilize the variances, and make the errors more normal.

1.13   BE FLEXIBLE ABOUT SCALE OF MEASUREMENT DETERMINING ANALYSIS

Introduction

What statistical methods are appropriate for what scale of measurement? This is a much debated topic. For example, is it appropriate to carry out an analysis of variance on a binomial variable?

Rule of Thumb

Do not let the scale of measurement rigidly determine the method of analysis.

Illustration

In some areas of research there is a somewhat dogmatic prescription that counting data should only be analyzed by chi-square statistics such as associated with contingency tables. This is overly strict. In most cases a reasonable analysis can be carried out on the means and variances of counting data. There are similar rules for other scales of measurement.

Basis of the Rule

The basis of the rule is that most of the time variables are mixtures or sums of diverse scales of measurement.

Discussion and Extensions

The classic statement about unit of measurement and statistical analysis is by Stevens (1946). The behavioral sciences literature has tended to codify this approach. However, there is still debate about these prescriptions. See, for example, Velleman and Wilkinson (1993) and Sarle (1996). Stevens categorized variables into counting, ordinal, interval, and ratio variables. Examples of each of these are binomial variables, rank order preferences, temperature scales, and scales of weight, respectively. Stevens then recommended that chi-square-like tests be used for binomial type variables, nonparametric tests be used for ordinal variables and that only for interval and ratio variables could parametric tests such as analysis of variance be used. While there is some common sense in these recommendations, they are unnecessarily restrictive and unrealistic. Furthermore, statistical practice has made them obsolete. For example, logistic regression methods combine binomial-like dependent variables with both discrete and continuous predictor variables. What all these prescriptions ignore is the asymptotic nature of many statistical procedures. The scale of measurement is less important than the distribution of values of the scale.

1.14   BE ECLECTIC AND ECUMENICAL IN INFERENCE

Introduction

There are several schools of thought about the nature of statistical inference. A primary difference among these schools is the view of probability: as a relative frequency or as a subjective assessment of the state of the world. At present, these two views appear to be irreconcilable. However, it may well be that like the particle-wave duality of light, these two approaches will ultimately be shown to be complementary. The most prominent schools of statistical inference are Neyman–Pearson, Likelihood, and Bayesian. These approaches are more similar than they are different. The first two use the relative frequency interpretation of probability, whereas the last one uses the subjective one.

Rule of Thumb

The practical applied statistician uses methods developed by all three schools as appropriate.

Illustration

Samet et al. (2000) used a variety of statistical procedures to estimate the health effects of air pollution. Bayesian and frequentist methods were used as thought appropriate; frequently the use of a method was based on the availability of statistical methodology to address the problem at hand.

Basis of the Rule

The basis of the rule is current statistical practice. In the majority of cases, statisticians use whatever approach makes sense and appears to be appropriate. Another reason is that in many cases the numerical results are identical, or at least similar, and it's the interpretation that differs.

Discussion and Extensions

The vast majority of statistical applications are valid in any of the schools. Quarrels about these topics among statisticians should be lovers' quarrels. It is a sign of statistical immaturity to denigrate any one of the schools. Oakes (1990) has written an excellent introduction to statistical inference. The three schools are described fairly and comprehensively. Oakes is very critical of tests of significance; in fact, he recommends against their use. He writes in the context of the social sciences, but his comments are also applicable to the natural sciences.

Discussions about the nature of statistical inference are no substitute for doing inference—that is, the practice of statistics. The philosophical arguments are important but they should not lead to apathy.

1.15 CONSIDER BOOTSTRAPPING FOR COMPLEX RELATIONSHIPS

Introduction

Bootstrapping, a procedure and term coined by Efron (1979), has become one of the most used statistical estimation procedures. Its popularity is due to a combination of available computing power and theoretical development. The challenge was to calculate the average, or expected value, of a complicated function of observed values of a variable. For example, the expected value of the square root of a Poisson random variable. A more traditional approach to estimating this quantity has been the delta method which involves a Taylor expansion of the square root function, and using the first few terms in the expansion. A great deal of theoretical statistics has involved developing and assessing such approximations.

Rule of Thumb

Think of bootstrapping instead of the delta method in estimating complex relationships.

Illustration

Efron (2002) provides a simple example involving the measurement of CD4 (Cluster of Differentiation on T4 cells) counts of 20 AIDS (Acquired Immunodeficiency Syndrome) patients before and after receiving an experimental drug. The Pearson correlation of the CD4 counts before and after treatment was 0.723. The challenge was to calculate a standard error (and confidence interval) associated with this value. Using the Fisher z-transformation (which is based on an application of the delta method), and a nonparametric method of estimation, led to a standard error of 0.0795. The bootstrap method produced a standard error of 0.0921, very comparable, requiring less mathematics, and fewer assumptions.

Basis of the Rule

There are two reasons for the rule. First, the delta method is an approximation which may not work well with small samples. Second, a sample from a population provides a crude estimate of the population distribution. Together with modern computing power, a sampling distribution can be generated and statistics estimated.

Discussion and Extensions

The bootstrap has become an important tool in estimation. Efron (2002) mentions more than 1000 papers on the theoretical basis of the bootstrap since 1979! One advantage of the bootstrap is that the technique can be applied to any data set without having to assume an underlying distribution. The delta method discussed in connection with the correlation coefficient still assumes that the transformed variate has a normal distribution. Additional comments on the delta method can be found in Rule 5.16 on page 125.

Some computer packages are beginning to provide bootstrap procedures. However, there is a need for a wider use and application of the technique. Most computer packages can handle very large numbers of repeated samplings, and this should not limit the accuracy of the bootstrap estimates.

A recent reference for bootstrap methods is the book by Davison and Hinkley (1997). There is also a discussion in this book of situations where the bootstrap does not work well, particularly in small sample situations where estimates may not exist. Nevertheless, the bootstrap has become a standard tool in the statisticians armamentarium and needs to be considered when complex estimation procedures are of interest.

1.16 ESTIMATE THE STANDARD ERROR WITH SAMPLE RANGE/SAMPLE SIZE

Introduction

It is frequently helpful to be able to estimate quickly the variability in a small data set. This can be useful for discussions of sample size and also to get some idea about the distribution of data.

Rule of Thumb

The standard error of a small sample of observations (15 observations or fewer) can be estimated by dividing the range by the number of observations in the sample,

$$ SE \approx \frac{Range}{n}, \tag{1.10} $$

where SE is the standard error, and n is the number of observations in the sample.

Table 1.5   Divisor for Sample Range to Estimate the Standard Error of the Sample Mean

Sample Size	Normal Distribution	Rectangular Distribution
2	1.6	1.6
3	2.9	3.0
5	5.2	5.2
10	9.7	9.0
15	13.4	11.7
20	16.7	14.0
50	31.8	23.5

Illustration

Consider the following sample of eight observations: 44, 48, 52, 60, 61, 63, 66, 69. The range, $69 - 44 = 25$, divided by n is $25/8 = 3.12$. The calculated standard error is 3.15.

Basis of the Rule

This rule is based on the distribution of the range of a random variable. In the case of the normal distribution there is no closed formula. The expected range of a random sample of size n from a standard normal distribution can be calculated from a table of order statistics for the standard normal distribution; the range is estimated by twice the largest expected value, $2E(X_n)$, where X_n is the largest expected value in a sample of size n. These values can be found in many tables: for example, Table A-21 in Fisher and van Belle (1993) lists the order statistics for a normal random variable. The expected value of the range for a normal distribution with standard deviation σ is then $2E(X_n)\sigma$. The above rule is based on the observation that the expected value of the range divided by the sample size is approximately equal to the standard error. This approximation is good for sample sizes of 15 or fewer and holds reasonably well even for samples from a uniform distribution. Table 1.5 illustrates the approximation.

For the uniform distribution there is an explicit formula for the expected value of the range based on the result that, for a random sample of size n from a Uniform $[0,1]$ distribution, the kth order statistic has expected value $k/(n + 1)$. Hence the range has expected value $(n - 1)/(n + 1)$. The variance of the uniform random variable is $1/12$. Hence, the range divided by the standard error is $\sqrt{12n}(n - 1)/(n + 1)$. Values in the last column in table 1.5 are based on this formula.

Discussion and Extensions

The estimate of the sample standard deviation can be obtained simply by multiplying the estimated standard error by the square root of the sample size. Alternatively, dividing the range by the square root of the sample size produces the estimate of the sample standard deviation. This rule, first introduced by Mantel (1951) is based on his Table II. Another useful reference for a more general discussion is Gehan (1980).

These results are a good example of a statistical rule of thumb. It is easy to remember, is reasonably accurate for small samples, and when applied will give a quick idea about the variability in the data. The following chapters provide many other examples of such rules.

2

Sample Size

The first question faced by a statistical consultant, and frequently the last, is, "How many subjects (animals, units) do I need?" This usually results in exploring the size of the treatment effects the researcher has in mind and the variability of the observational units. Researchers are usually less interested in questions of Type I error, Type II error, and one-sided versus two-sided alternatives. A key question is to settle the type of variable (endpoint) the consultee has in mind: Is it continuous, discrete, or something else? For continuous measurements the normal distribution is the default model, for distributions with binary outcomes, the binomial.

The ingredients in a sample size calculation, for one or two groups, are:

Type I Error (α)	Probability of rejecting the null hypothesis when it is true
Type II Error (β)	Probability of not rejecting the null hypothesis when it is false
Power = $1 - \beta$	Probability of rejecting the null hypothesis when it is false
σ_0^2 and σ_1^2	Variances under the null and alternative hypotheses (may be the same)
μ_0 and μ_1	Means under the null and alternative hypotheses
n_0 and n_1	Sample sizes in two groups (may be the same)

The choice of the alternative hypothesis is challenging. Researchers sometimes say that if they knew the value of the alternative hypothesis, they would not need to do the study. There is also debate about which is the null hypothesis and which is the

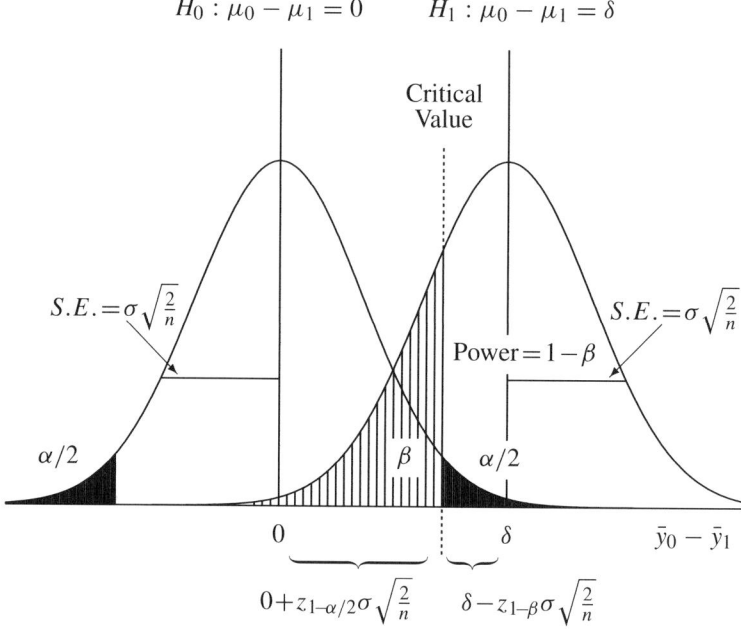

$$H_0 : \mu_0 - \mu_1 = 0 \qquad H_1 : \mu_0 - \mu_1 = \delta$$

Fig. 2.1 Sampling model for two independent sample case. Two-sided alternative, equal variances under null and alternative hypotheses.

alternative hypothesis. The discussion can become quite philosophical, but there are practical implications as well. In environmental studies does one assume that a site is safe or hazardous as the null hypothesis? Millard (1987a) argues persuasively that the choice affects sample size calculations. This is a difficult issue. Fortunately, in most research settings the null hypothesis is reasonably assumed to be the hypothesis of no effect. There is a need to become familiar with the research area in order to be of more than marginal use to the investigator. In terms of the alternative hypothesis, it is salutary to read the comments of Wright (1999) in a completely different context, but very applicable to the researcher: "an alternative hypothesis ... must make sense of the data, do so with an essential simplicity, and shed light on other areas." This provides some challenging guidance to the selection of an alternative hypothesis.

The phrase, "Type I error," is used loosely in the statistical literature. It can refer to the error as such, or the probability of making a Type I error. It will usually be clear from the context which is meant.

Figure 2.1 summarizes graphically the ingredients in sample size calculations. The null hypothesis provides the basis for determining the rejection region, whether the test is one-sided or two-sided, and the probability of a Type I error (α)—the size of the test. The alternative hypothesis then defines the power and the Type II error (β). Notice that moving the curve associated with the alternative hypothesis to the right (equivalent to increasing the distance between null and alternative hypotheses) increases the area of the curve over the rejection region and thus increases the power. The *critical value* defines the boundary between the rejection and nonrejection regions. This value must be the same under the null and alternative hypotheses. This then leads to the fundamental equation for the two-sample situation:

$$0 + z_{1-\alpha/2}\sigma\sqrt{\frac{2}{n}} = \delta - z_{1-\beta}\sigma\sqrt{\frac{2}{n}}. \tag{2.1}$$

If the variances, and sample sizes, are not equal, then the standard deviations in equation (2.1) are replaced by the values associated with the null and alternative hypotheses, and individual sample sizes are inserted as follows,

$$0 + z_{1-\alpha/2}\sigma_0\sqrt{\frac{1}{n_0} + \frac{1}{n_1}} = \delta - z_{1-\beta}\sqrt{\frac{\sigma_0^2}{n_0} + \frac{\sigma_1^2}{n_1}}. \tag{2.2}$$

This formulation is the most general and is the basis for virtually all two-sample sample size calculations. These formulae can also be used in one-sample situations by assuming that one of the samples has an infinite number of observations.

2.1 BEGIN WITH A BASIC FORMULA FOR SAMPLE SIZE

Introduction

Start with the basic sample size formula for two groups, with a two-sided alternative, normal distribution with homogeneous variances ($\sigma_0^2 = \sigma_1^2 = \sigma^2$) and equal sample sizes ($n_0 = n_1 = n$).

Rule of Thumb

The basic formula is

$$n = \frac{16}{\Delta^2}, \tag{2.3}$$

where

$$\Delta = \frac{\mu_0 - \mu_1}{\sigma} = \frac{\delta}{\sigma} \tag{2.4}$$

is the treatment difference to be detected in units of the standard deviation—the standardized difference.

In the one-sample case the numerator is 8 instead of 16. This situation occurs when a single sample is compared with a known population value.

Illustration

If the standardized difference, Δ, is expected to be 0.5, then $16/0.5^2 = 64$ subjects per treatment will be needed. If the study requires only one group, then a total

Table 2.1 Numerator for Sample Size Formula, Equation (2.3); Two-Sided Alternative Hypothesis, Type I Error, $\alpha = 0.05$

Type II Error β	Power $1 - \beta$ Power	Numerator for Sample Size Equation (2.3)	
		One Sample	Two Sample
0.50	0.50	4	8
0.20	0.80	8	16
0.10	0.90	11	21
0.05	0.95	13	26
0.025	0.975	16	31

of 32 subjects will be needed. The two-sample scenario will require 128 subjects, the one-sample scenario one-fourth of that number. This illustrates the rule that the two-sample scenario requires four times as many observations as the one-sample scenario. The reason is that in the two-sample situation two means have to be estimated, doubling the variance, and, additionally, requires two groups.

Basis of the Rule

The formula for the sample size required to compare two population means, μ_0 and μ_1, with common variance, σ^2, is

$$n = \frac{2\left(z_{1-\alpha/2} + z_{1-\beta}\right)^2}{\left(\dfrac{\mu_0 - \mu_1}{\sigma}\right)^2}. \tag{2.5}$$

This equation is derived from equation (2.1). For $\alpha = 0.05$ and $\beta = 0.20$ the values of $z_{1-\alpha/2}$ and $z_{1-\beta}$ are 1.96 and 0.84, respectively; and $2(z_{1-\alpha/2} + z_{1-\beta})^2 = 15.68$, which can be rounded up to 16, producing the rule of thumb above.

Discussion and Extensions

This rule—which should be memorized—can also be found in Snedecor and Cochran (1989).

The two key ingredients are the difference to be detected, $\delta = \mu_0 - \mu_1$, and the inherent variability of the observations indicated by σ^2. The numerator can be calculated for other values of Type I and Type II error. Table 2.1 lists the values of the numerator for Type I error of 0.05 and different values of Type II error and power. A power of 0.90 or 0.95 is frequently used to evaluate new drugs in Phase III clinical trials (usually double blind comparisons of new drug with placebo or standard); see

Lakatos (1998). One advantage of a power of 0.95 is that it bases the inferences on confidence intervals.

The two most common sample size situations involve one or two samples. Since the numerator in the rule of thumb is 8 for the one-sample case, this illustrates that the two-sample situation requires four times as many observations as the one-sample case. This pattern is confirmed by the numerators for sample sizes in Table 2.1.

If the researcher does not know the variability and cannot be led to an estimate, the discussion of sample size will have to be addressed in terms of standardized units. A lack of knowledge about variability of the measurements indicates that substantial education is necessary before sample sizes can be calculated.

Equation (2.3) can be used to calculate detectable difference for a given sample size, n. Inverting this equation gives

$$\Delta = \frac{4}{\sqrt{n}}, \tag{2.6}$$

or

$$\mu_0 - \mu_1 = \frac{4\sigma}{\sqrt{n}}. \tag{2.7}$$

In words, the detectable *standardized* difference in the two-sample case is about 4 divided by the square root of the number of observations per sample. The detectable (non-standardized) difference is four standard deviations divided by the square root of the number of observations per sample. For the one-sample case the numerator 4 is replaced by 2, and the equation is interpreted as the detectable deviation from some parameter value μ. Figure 2.2 relates sample size to power and detectable differences for the case of Type I error of 0.05. This figure also can be used for estimating sample sizes in connection with correlation, as discussed in Rule 3.4 on page (59).

This rule of thumb, represented by equation (2.2), is very robust and useful for sample size calculations. Many sample size questions can be formulated so that this rule can be applied.

2.2 IGNORE THE FINITE POPULATION CORRECTION IN CALCULATING SAMPLE SIZE FOR A SURVEY

Introduction

Survey sampling questions are frequently addressed in terms of wanting to know a population proportion with a specified degree of precision. The sample size formula can be used with a power of 0.5 (which makes $z_{1-\beta} = 0$). The numerator for the two-sample situation then becomes 8, and 4 for the one sample situation (from Table 2.1).

Survey sampling typically deals with a finite population of size N with a corresponding reduction in the variability if sampling is without replacement. The reduction in the standard deviation is known as the finite population correction. Specifically, a sample of size n is taken without replacement from a population of size N, and the

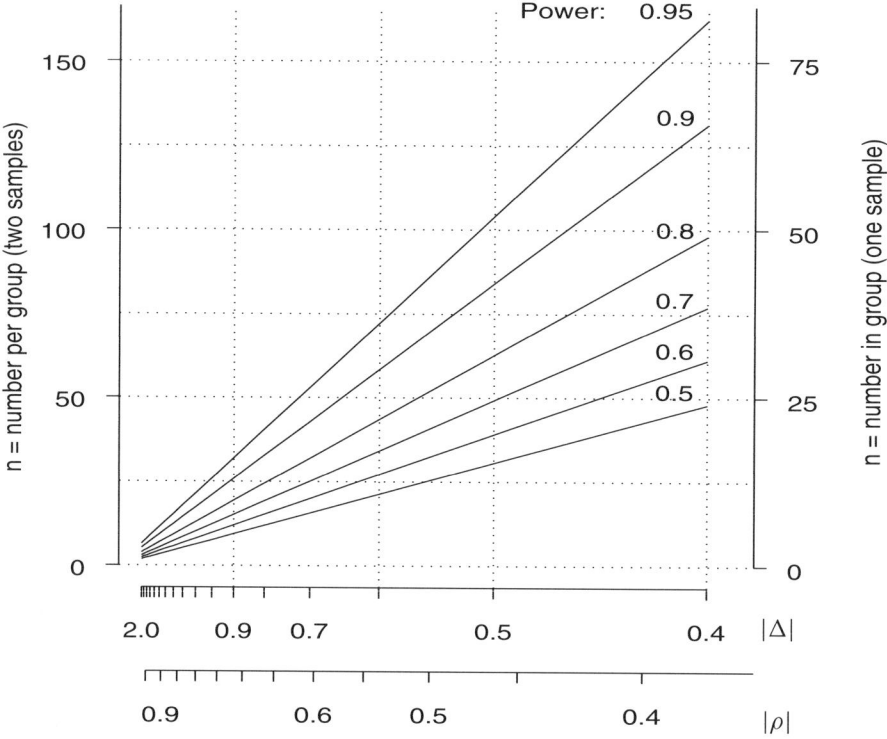

Fig. 2.2   Sample size for one sample and two samples. Use right-hand side for one-sample situation and correlation.

sample mean and its standard error are calculated. Then the standard error of the sample mean, $\bar{x}$ is

$$SE(\bar{x}) = \sqrt{\frac{N-n}{nN}}\,\sigma.$$ (2.8)

Rule of Thumb

The finite population correction can be ignored in initial discussions of survey sample size questions.

Illustration

A sample of 50 is taken without replacement from a population of 1000. Assume that the standard deviation of the population is 1. Then the standard error of the mean ignoring the finite population correction is 0.141. Including the finite population correction leads to a standard error of 0.138.

Basis of the Rule

The standard error with the finite population correction can be written as

$$SE(\bar{x}) = \frac{1}{\sqrt{n}}\sqrt{1 - \frac{n}{N}}\sigma. \tag{2.9}$$

So the finite population correction, $\sqrt{1 - \frac{n}{N}}$, is a number less than one, and the square root operation pulls it closer to one. If the sample is 10% of the population, the finite population correction is 0.95 or there is a 5% reduction in the standard deviation. This is not likely to be important and can be ignored in most preliminary reviews.

Discussion and Extensions

If the population is very large, as is frequently the case, the finite population correction can be ignored throughout the inference process. The formula also indicates that changes in orders of magnitude of the population will not greatly affect the precision of the estimate of the mean.

The rule indicates that initial sample size calculations can ignore the finite population correction, and the precision of the estimate of the population mean is proportional to $\sqrt{n}$ (for fixed standard deviation, σ).

2.3 CALCULATING SAMPLE SIZE USING THE COEFFICIENT OF VARIATION

Introduction

Consider the following dialogue in a consulting session,

"What kind of treatment effect are you anticipating?"

"Oh, I'm looking for a 20% change in the mean."

"Mm, and how much variability is there in your observations?"

"About 30%"

The dialogue indicates how researchers frequently think about relative treatment effects and variability. How to address this question? It turns out, fortuitously, that the question can be answered.

Rule of Thumb

The sample size formula becomes:

$$n = \frac{8(CV)^2}{(PC)^2}[1 + (1 - PC)^2]. \tag{2.10}$$

where PC is the proportionate change in means ($PC = (\mu_0 - \mu_1)/\mu_0$) and CV is the coefficient of variation ($CV = \sigma_0/\mu_0 = \sigma_1/\mu_1$).

Table 2.2 Sample Sizes for a Range of Coefficients of Variation and Percentage Change in Means: Two Sample Tests, Two-sided Alternatives with Type I Error 0.05 and Power 0.80 (Calculations Based on Equation 2.10)

		Percentage Change in Means						
		5	**10**	**15**	**20**	**30**	**40**	**50**
	5	16	4	2	1	1	1	1
	10	61	15	7	4	2	1	1
Coefficient	**15**	137	33	14	8	3	2	1
of	**20**	244	58	25	14	6	3	2
Variation	**30**	548	131	56	30	12	7	4
in	**40**	975	232	98	53	22	11	7
Percent	**50**	>1000	362	154	82	34	17	10
	75	>1000	815	345	185	75	39	23
	100	>1000	>1000	613	328	133	68	40

Illustration

For the situation described in the consulting session the sample size becomes,

$$n = \frac{8(0.30)^2}{(0.20)^2} [1 + (1 - 0.20)^2].$$

$$= 29.52 \approx 30$$

and the researcher will need to aim for about 30 subjects per group. If the treatment is to be compared with a standard, that is, only one group is needed, then the sample size required will be 15.

Basis of the Rule

Since the coefficient of variation is assumed to be constant this implies that the variances of the two populations are not the same and the variance σ^2 in equation (2.5) is replaced by the average of the two population variances:

$$\sigma^2 = \frac{\sigma_0^2 + \sigma_1^2}{2}. \tag{2.11}$$

Replacing σ_i by $\mu_i CV$ for $i = 0, 1$ and simplifying the algebra leads to the equation above.

Table 2.2 lists the sample sizes based on equation (2.10) for values of CV ranging from 5% to 100% and values of PC ranging from 5% to 50%. These are ranges most likely to be encountered in practice.

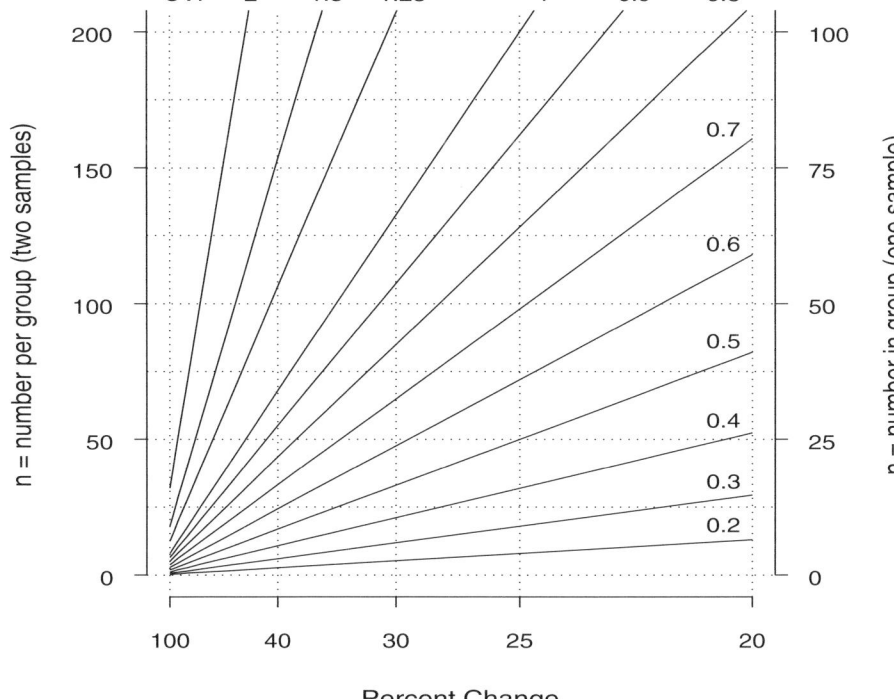

Fig. 2.3 Sample size for coefficient of variation and percent change. Conditions the same as for Table 2.2.

Discussion and Extensions

Figure 2.3 presents an alternative way to estimating sample sizes for the situation where the specifications are made in terms of percentage change and coefficient of variation. For the data of the illustration the same sample size is obtained by using this figure.

Sometimes the researcher will not have any idea of the variability inherent in the system. For biological variables, a variability on the order of 35% is not uncommon, and will form the basis for a discussion by assuming a sample size formula of:

$$n \simeq \frac{1}{(PC)^2}[1 + (1 - PC)^2].\qquad(2.12)$$

This formula can be simplified further for relatively small values of percentage change by noting that $1 - PC \approx 1$ and a simple formula to keep in mind is,

$$n = \frac{2}{(PC)^2}.\qquad(2.13)$$

For additional discussion see van Belle and Martin (1993).

2.4   DO NOT FORMULATE A STUDY SOLELY IN TERMS OF EFFECT SIZE

Introduction

The standardized difference, Δ, is sometimes called the *effect size*. As discussed in Rule 2.1, it scales the difference in population means by the standard deviation. Figure 2.1 is based on this standardized difference. While this represents a useful approach to sample size calculations, a caution is in order. The caution is serious enough that it merits a separate rule of thumb.

Rule of Thumb

Do not formulate objectives for a study solely in terms of effect size.

Illustration

Some social science journals insist that all hypotheses be formulated in terms of effect size. This is an unwarranted demand that places research in an unnecessary straightjacket.

Basis of the Rule

There are two reasons for caution in the use of effect size. First, the effect size is a function of at least three parameters, representing a substantial reduction of the parameter space. Second, the experimental design may preclude estimation of some of the parameters.

Discussion and Extensions

Insistence on effect size as a non-negotiable omnibus quantity (see Rule 1.3) forces the experimental design, or forces the researcher to get estimates of parameters from the literature. For example, if effect size is defined in terms of a subject-subject variance, then a design involving paired data will not be able to estimate that variance. So the researcher must go elsewhere to get the estimate, or change the design of the study to get it. This is unnecessarily restrictive.

The use of Δ in estimating sample size is not restrictive. For a given standard deviation the needed sample size may be too large and the researcher may want to look at alternative designs for possible smaller variances, including the use of covariates.

An omnibus quantity simplifies a complicated parameter or variable space. This is desirable in order to get a handle on a research question. It does run the danger of making things too simple (Rule 1.5). One way to protect against this danger is to use effect size as one of several means of summarizing the data.

2.5 OVERLAPPING CONFIDENCE INTERVALS DO NOT IMPLY NONSIGNIFICANCE

Introduction

It is sometimes claimed that if two independent statistics have overlapping confidence intervals, then they are not significantly different. This is certainly true if there is substantial overlap. However, the overlap can be surprisingly large and the means still significantly different.

Rule of Thumb

Confidence intervals associated with statistics can overlap as much as 29% and the statistics can still be significantly different.

Illustration

Consider means of two independent samples. Suppose their values are 10 and 22 with equal standard errors of 4. The 95% confidence intervals for the two statistics, using the critical value of 1.96, are 2.2–17.8 and 14.2–29.8, displaying considerable overlap. However, the z-statistic comparing the two means has value

$$z = \frac{22 - 10}{\sqrt{4^2 + 4^2}} = 2.12.$$

Using the same criterion as applied to the confidence intervals this result is clearly significant. Analogously, the 95% confidence interval for the difference is $(22 - 10) \pm 1.96\sqrt{4^2 + 4^2}$ producing the interval 0.9–23.1. This interval does not straddle 0 and the conclusion is the same.

Basis of the Rule

Assume a two-sample situation with standard errors equal to σ_1 and σ_2 (this notation is chosen for convenience). For this discussion assume that a 95 % confidence level is of interest. Let the multiplier for the confidence interval for each mean be k to be chosen so that the mean difference just reaches statistical significance. The following equation must then hold,

$$\frac{k\sigma_1 + k\sigma_2}{\sqrt{\sigma_1^2 + \sigma_2^2}} = 1.96. \tag{2.14}$$

Assuming that $\sigma_1 = \sigma_2$ leads to the result $k = 1.39$. That is, if the half-confidence interval is (1.39 × standard error) the means will be significantly different at the 5% level. Thus, the overlap can be $1 - 1.39/1.96 \sim 29\%$.

Discussion and Extensions

If the standard errors are not equal—due to heterogeneity of variances or unequal sample sizes—the overlap, which maintains a significant difference, decreases. Specifically, for a 95% confidence interval situation, if $r = \sigma_1/\sigma_2$ then

$$k = \sqrt{1 - \frac{2r}{(r+1)^2}}\; 1.96. \tag{2.15}$$

This shows that the best that can be done is when $r = 1$. As r moves away from 1 (either way), the correction approaches 1, and k approaches 1.96.

The multiplier of 1.96 does not depend on the confidence level so the corrections apply regardless of the level. Another feature is that the correction involves a square root so is reasonably robust. For $r = 2$, corresponding to a difference in variances of 4, the overlap can still be 25%.

This rules indicates that it is insufficient to only consider non-overlapping confidence intervals to represent significant differences between the statistics involved. It may require a quick calculation to establish significance or non-significance. A good rule of thumb is to assume that overlaps of 25% or less still suggest statistical significance.

2.6 SAMPLE SIZE CALCULATION FOR THE POISSON DISTRIBUTION

Introduction

The Poisson distribution is known as the law of small numbers, meaning that it deals with rare events. The term "rare" is undefined and needs to be considered in context. A rather elegant result for sample size calculations can be derived in the case of Poisson variables. It is based on the square root transformation of Poisson random variables.

Rule of Thumb

Suppose the means of samples from two Poisson populations are to be compared in a two-sample test. Let θ_0 and θ_1 be the means of the two populations. Then the required number of observations per sample is

$$n = \frac{4}{(\sqrt{\theta_0} - \sqrt{\theta_1})^2}. \tag{2.16}$$

Illustration

Suppose the hypothesized means are 30 and 36. Then the number of sampling units per group is required to be $4/(\sqrt{30} - \sqrt{36})^2 = 14.6 = 15$ observations per group.

Basis of the Rule

Let Y_i be Poisson with mean θ_i for $i = 0, 1$. Then it is known that $\sqrt{Y_i}$ is approximately normal ($\mu_i = \sqrt{\theta_i}$, $\sigma^2 = 0.25$). Using equation (2.3) the sample size formula for the Poisson case becomes equation (2.16).

Discussion and Extensions

The sample size formula can be rewritten as

$$n = \frac{4}{(\theta_0 + \theta_1)/2 - \sqrt{\theta_0 \theta_1}}. \tag{2.17}$$

The denominator is the difference between the arithmetic and the geometric means of the two Poisson distributions! The denominator is always positive since the arithmetic mean is larger than the geometric mean (Jensen's inequality).

Now suppose that the means θ_0 and θ_1 are means per unit time (or unit volume) and that the observations are observed for a period of time, T. Then Y_i is Poisson with mean $\theta_i T$. Hence, the sample size required is

$$n = \frac{4}{T(\sqrt{\theta_0} - \sqrt{\theta_1})^2}. \tag{2.18}$$

This formula is worth contemplating. Increasing the observation period T, reduces the sample size proportionately, not as the square root! This is a basis for the observation that the precision of measurements of radioactive sources, which often follow a Poisson distribution, can be increased by increasing the duration of observation times.

Choose T so that the number per sample is 1. To achieve that effect choose T to be of length

$$T = \frac{4}{(\sqrt{\theta_0} - \sqrt{\theta_1})^2}. \tag{2.19}$$

This again is reasonable since the sum of independent Poisson variables is Poisson, that is, ΣY_i is Poisson ($T\theta_i$) if each Y_i is Poisson (θ_i). This formulation will be used in Chapter 4, Rule 4.3 page 82, which discusses the number of events needed in the context of epidemiological studies.

The Poisson distribution can be considered the basis for a large number of discrete distributions such as the binomial, hypergeometric, and multinomial distributions. It is important to be familiar with some of these basic properties.

2.7 SAMPLE SIZE CALCULATION FOR POISSON DISTRIBUTION WITH BACKGROUND RATE

Introduction

The Poisson distribution is a common model for describing radioactive scenarios. Frequently there is background radiation over and above which a signal is to be detected. Another application is in epidemiology when a disease has a low background rate and a risk factor increases that rate. This is true for most disease situations. It turns out that the higher the background rate the larger the sample size needed to detect differences between two groups.

Rule of Thumb

Suppose that the background rate is θ^* and let θ_0 and θ_1 now be the additional rates over background. Then, Y_i is Poisson $(\theta^* + \theta_i)$. The rule of thumb sample size formula is

$$n = \frac{4}{(\sqrt{\theta^* + \theta_0} - \sqrt{\theta^* + \theta_1})^2}. \tag{2.20}$$

Illustration

Suppose the means of the two populations of radioactive counts are 1 and 2 with no background radiation. Then the sample size per group to detect this difference, using equation (2.16), is $n = 24$. Now assume a background level of radiation of 1.5. Then the sample size per group, using equation (2.20), becomes 48. Thus the sample size has doubled with a background radiation halfway between the two means.

Basis of the Rule

The result follows directly by substituting the means $(\theta^* + \theta_0)$ and $(\theta^* + \theta_1)$ into equation (2.16).

Discussion and Extensions

It has been argued that older people survive until an important event in their lives has occurred and then die. For example, the number of deaths in the United States in the week before the year 2000 should be significantly lower than the number of deaths in the week following (or, perhaps better, the third week before the new year and the third week after the new year). How many additional deaths would need to be observed to be fairly certain of picking up some specified difference? This can be answered by equation (2.20). For this particular case assume that $\theta_0 = 0$ and $\theta_1 = \Delta\theta$. That is, $\Delta\theta$ is the increase in the number of deaths needed to have a power of 0.80 that it will be picked up, if it occurs. Assume also that the test is two-sided—there could be a decrease in the number of deaths. Suppose the average number of deaths per week in the United States (θ^*) is 50,000—a reasonable number. The sample size is $n = 1$. Some manipulation of equation (2.7) produces

$$\Delta\theta = 4\sqrt{\theta^*}. \tag{2.21}$$

For this situation an additional number of deaths equal to $4 \times \sqrt{50,000} = 894.4 \sim 895$ would be needed to be reasonably certain that the assertion had been validated. All these calculations assume that the weeks were pre-specified without looking at the data. Equation (2.21) is very useful in making a quick determination about increases in rates that can be detected for a given background rate. In the one-sample situation the multiplier 4 can be replaced by 2.

A second application uses equation (2.3) as follows. Suppose n^* is the sample size associated with the situation of a background rate of θ^*. Let $\bar{\theta} = (\theta_0 + \theta_1)/2$ be the arithmetic mean of the two parameters. Then using equation (2.3) for the sample size calculation (rather than the square root formulation) it can be shown that

$$n^* = n \left(1 + \frac{\theta^*}{\bar{\theta}} \right). \tag{2.22}$$

Thus, if the background rate is twice the average increase to be detected, then the sample size is doubled. This confirms the calculation of the illustration. The problem could also have been formulated by assuming the background rate increased by a factor R so that the rates are θ^* and $R\theta^*$. This idea will be explored in Chapter 4, Rule 4.3.

This rule of thumb is a good illustration of how a basic rule can be modified in a straightforward way to cover more general situations of surprising usefulness.

2.8   SAMPLE SIZE CALCULATION FOR THE BINOMIAL DISTRIBUTION

Introduction

The binomial distribution provides a useful model for independent Bernoulli trials. The sample size formula in equation (2.3) can be used for an approximation to the sample size question involving two independent binomial samples. Using the same labels for variables as in the Poisson case, let Y_i, $i = 0, 1$ be independent binomial random variables with probability of success π_i, respectively. Assume that equal sample sizes, n, are required.

Rule of Thumb

To compare two proportions, π_0 and π_1 use the formula

$$n = \frac{16\bar{\pi}(1 - \bar{\pi})}{(\pi_0 - \pi_1)^2}, \tag{2.23}$$

where $\bar{\pi} = (\pi_0 + \pi_1)/2$ is used to calculate the average variance.

Illustration

For $\pi_0 = 0.3$ and $\pi_1 = 0.1$, $\bar{\pi} = 0.2$ so that the required sample size per group is $n = 64$.

Basis of the Rule

Use equation (2.3) with the variance estimated by $\bar{\pi}(1 - \bar{\pi})$.

Discussion and Extensions

Some care should be taken with this approximation. It is reasonably good for values of n that come out between 10 and 100. For larger (or smaller) resulting sample sizes using this approximation, more exact formulae should be used. For more extreme values, use tables of exact values given by Haseman (1978) or use more exact formulae (see Fisher and van Belle, 1993). Note that the tables by Haseman are for one-tailed tests of the hypotheses, thus they will tend to be smaller than sample sizes based on the two-tailed assumption in equation (2.3).

An upper limit on the required sample size is obtained using the maximum variance of 1/4 which occurs at $\pi_i = 1/2$. For these values $\sigma = 1/2$ and the sample size formula becomes

$$n = \frac{4}{(\pi_0 - \pi_1)^2}. \tag{2.24}$$

This formula produces a conservative estimate of the sample size. Using the specification in the illustration produces a sample size of $n = 4/(0.3 - 0.1)^2 = 100$—considerably higher than the value of 64. This is due to the larger value for the variance. This formula is going to work reasonably well when the proportions are centered around 0.5.

Why not use the variance stabilizing transformation for the binomial case? This has been done extensively. The variance stabilizing transformation for a binomial random variable, X, the number of successes in n Bernoulli trials with probability of success, π is,

$$Y = \sin^{-1}\sqrt{\frac{X}{n}}, \tag{2.25}$$

where the angle is measured in radians. The variance of $Y = 1/(4n)$. Using the arcsine transformation in equation (2.3) gives

$$n = \frac{4}{\left(\sin^{-1}\sqrt{\pi_0} - \sin^{-1}\sqrt{\pi_1}\right)^2}. \tag{2.26}$$

For the example this produces $n = 60.1 = 61$; the value of n=64, using the more easily remembered equation (2.23) is compatible.

For proportions less than 0.05, $\sin^{-1}\sqrt{\pi} \approx \sqrt{\pi}$. This leads to the sample size formula,

$$n = \frac{4}{(\sqrt{\pi_0} - \sqrt{\pi_1})^2}, \tag{2.27}$$

which is linked to the Poisson formulation in equation (2.16).

Equation (2.1) assumes that the variances are equal or can be replaced by the average variance. Many reference books, for example Selvin (1996) and Lachin (2000), use equation (2.2) with the variances explicitly accounted for. The hypothesis testing situation is, $H_0 : \pi_0 = \pi_1 = \pi$ and $H_1 : \pi_0 \neq \pi_1$, say, $\pi_0 - \pi_1 = \delta$. This produces the fundamental equation

$$0 + z_{1-\alpha/2}\sqrt{\frac{2\pi(1-\pi)}{n}} = \delta - z_{1-\beta}\sqrt{\frac{\pi_0(1-\pi_0)}{n} + \frac{\pi_1(1-\pi_1)}{n}}. \qquad (2.28)$$

Solving this equation for n produces

$$n = \frac{\left(z_{1-\alpha/2}\sqrt{2\pi(1-\pi)} + z_{1-\beta}\sqrt{\pi_0(1-\pi_0) + \pi_1(1-\pi_1)}\right)^2}{(\pi_0 - \pi_1)^2}. \qquad (2.29)$$

Using this equation (2.29) for the data in the illustration with $\pi_0 = \pi_1 = 0.2$ under the null hypothesis, and $\pi_0 = 0.3$, $\pi_1 = 0.1$ under the alternative hypothesis produces a sample size of $n = 61.5 \sim 62$. Clearly, the approximation works very well.

As illustrated, equation (2.23) produces reasonable estimates of sample sizes for n in the range from 10 to 100. For smaller sampling situations exact tables should be used.

2.9 WHEN UNEQUAL SAMPLE SIZES MATTER; WHEN THEY DON'T

Introduction

In some cases it may be useful to have unequal sample sizes. For example, in epidemiological studies it may not be possible to get more cases, but more controls are available. Suppose n subjects are required per group, but only n_0 are available for one of the groups, assuming that $n_0 < n$. What is the number of subjects, kn_0, required in the second group in order to obtain the same precision as with n in each group?

Rule of Thumb

To get equal precision with a two-sample situation with n observations per sample given n_0 ($n_0 < n$) in the first sample and kn_0 observations in the second sample, choose

$$k = \frac{n}{2n_0 - n}. \qquad (2.30)$$

Illustration

Suppose that sample size calculations indicate that $n = 16$ cases and controls are needed in a case-control study. However, only 12 cases are available. How many controls will be needed to obtain the same precision? The answer is $k = 16/8 = 2$ so that 24 controls will be needed to obtain the same precision as with 16 cases and controls.

Basis of the Rule

For two independent samples of size n, the variance of the estimate of difference (assuming equal variances) is proportional to

$$\frac{1}{n} + \frac{1}{n} = \frac{2}{n}. \tag{2.31}$$

Given a sample size $n_0 < n$ available for the first sample and a sample size kn_0 for the second sample and then equating the variances for the two designs, produces

$$\frac{1}{n_0} + \frac{1}{kn_0} = \frac{2}{n}. \tag{2.32}$$

Solving for k produces the result.

Discussion and Extensions

The rule of thumb implies that there is a lower bound to the number of observations in the smaller sample size group. In the example the required precision, as measured by the sum of reciprocals of the sample sizes is $1/8$. Assuming that $k = \infty$ requires 8 observations in the first group. This is the minimum number of observations. Another way of saying the result is that it is not possible to reduce the variance of the difference to less than $1/8$. This result is asymptotic. How quickly is the minimum value of the variance of the difference approached? This turns out the be a function of k only.

   This approach can be generalized to situations where the variances are not equal. The derivations are simplest when one variance is fixed and the second variance is considered a multiple of the first variance (analogous to the sample size calculation).

   Now consider two designs, one with n observations in each group and the other with n and kn observations in each group. The relative precision of these two designs is

$$\frac{SE_k}{SE_0} = \sqrt{\frac{1}{2}\left(1 + \frac{1}{k}\right)}, \tag{2.33}$$

where SE_k and SE_0 are the standard errors of the designs with kn and n subjects in the two groups, respectively. Using $k = 1$, results in the usual two-sample situation with equal sample size. If $k = \infty$, the relative precision is $\sqrt{0.5} = 0.71$. Hence, the best that can be done is to decrease the standard error of the difference by 29%. For $k = 4$ the value is already 0.79 so that from the point of view of precision there is no reason to go beyond four or five times more subjects in the second group than the first group. This will come close to the maximum possible precision in each group.

   There is a converse to the above rule: minor deviations from equal sample sizes do not affect the precision materially. Returning to the illustration, suppose the sample size in one group is 17, and the other is 15 so that the total sampling effort is the same. In this case the precision is proportional to

$$\frac{1}{17} + \frac{1}{15} = 0.1255.$$

This compares with 0.125 under the equal sampling case. Thus the precision is virtually identical and some imbalance can be tolerated. Given that the total sampling effort remains fixed a surprisingly large imbalance can be tolerated. Specifically, if the samples are split into $0.8n$ and $1.2n$ the decrease in precision, as measured by the reciprocal of the sample sizes, in only 4%. So an imbalance of approximately 20% has a small effect on precision. A split of $0.5n$, $1.5n$ results in a 33% reduction in precision. The results are even more comparable if the precision is expressed in terms of standard errors rather than variances. It must be stressed that this is true only if the total sampling effort remains the same. Also, if the variances are not equal the results will differ, but the principle remains valid. See the next rule for further discussion. Lachin (2000) gives a good general approach to sample size calculations with unequal numbers of observations in the samples.

The importance of the issue of unequal sample sizes must be considered from two points of view: when unequal sample sizes matter and when they don't. It matters when multiple samples can be obtained in one group. It does not matter under moderate degrees of imbalance.

2.10 DETERMINING SAMPLE SIZE WHEN THERE ARE DIFFERENT COSTS ASSOCIATED WITH THE TWO SAMPLES

Introduction

In some two-sample situations the cost per observation is not equal and the challenge then is to choose the sample sizes in such a way so as to minimize cost and maximize precision, or minimize the standard error of the difference (or, equivalently, minimize the variance of the difference). Suppose the cost per observation in the first sample is c_0 and in the second sample is c_1. How should the two sample sizes n_0 and n_1 be chosen?

Rule of Thumb

To minimize the total cost of a study, choose the ratio of the sample sizes according to

$$\frac{n_1}{n_0} = \sqrt{\frac{c_0}{c_1}}. \tag{2.34}$$

This is the square root rule: Pick sample sizes inversely proportional to square root of the cost of the observations. If costs are not too different, then equal sample sizes are suggested (because the square root of the ratio will be closer to 1).

Illustration

Suppose the cost per observation for the first sample is 160 and the cost per observation for the second sample is 40. Then the rule of thumb recommends taking twice as many observations in the second group as compared to the first. To calculate the specific sample sizes, suppose that on an equal sample basis 16 observations are needed. To get equal precision with n_0 and $2n_0$, use equation (2.32) with $k = 2$ to produce 12 and 24 observations, respectively. The total cost is then $12 \times 160 + 24 \times 40 = 2800$ compared with the equal sample size cost of $16 \times 160 + 16 \times 40 = 3200$ for a 10% saving in total cost. This is a modest saving in total cost for a substantial difference in cost per observation in the two samples. This suggests that costs are not going to play a major role in sample size determinations.

Basis of the Rule

The cost, C, of the experiment is

$$C = c_0 n_0 + c_1 n_1, \tag{2.35}$$

where n_0 and n_1 are the number of observations in the two groups, respectively, and are to be chosen to minimize

$$\frac{1}{n_0} + \frac{1}{n_1}, \tag{2.36}$$

subject to the total cost being C. This is a linear programming problem with solutions:

$$n_0 = \frac{C}{c_0 + \sqrt{c_0 c_1}} \tag{2.37}$$

and

$$n_1 = \frac{C}{c_1 + \sqrt{c_0 c_1}}. \tag{2.38}$$

When ratios are taken, the result follows.

Discussion and Extensions

The argument is similar to that in connection with the unequal sample size rule of thumb. There are also two perspectives in terms of precision: when do costs matter, when do they not? The answer is in the same spirit as that associated with Rule 2.9. On the whole, costs are not that important. A little algebra shows that the total cost of a study under equal sample size, say C_{equal} is related to the total cost of a study using the square root rule, say $C_{optimal}$ as follows,

$$\frac{C_{equal} - C_{optimal}}{C_{equal}} = \frac{1}{2} - \frac{\sqrt{c_0 c_1}}{c_0 + c_1}. \tag{2.39}$$

The following array displays the savings as a function of the differential costs per observation in the two samples. It is assumed that the costs are higher in the first sample.

$\dfrac{c_0}{c_1}$	1	2	5	10	15	20	100
$\dfrac{C_{equal} - C_{optimal}}{C_{equal}}$	0	3%	13%	21%	26%	29%	40%.

These results are sobering. The savings can never be greater than 50%. Even a five-fold difference in cost per observation in the two groups results in only a 13% reduction in the total cost. These results are valid for all sample sizes, that is, the percentage savings is a function of the different costs per observation, not the sample sizes. A similar conclusion is arrived at in Rule 4.5 on page (86).

This discussion assumes—unrealistically—that there are no overhead costs. If overhead costs are taken into account, the cost per observation will change and, ordinarily, reduce the impact of cost on precision.

The variance of an observation can be considered a cost; the larger the variance the more observations are needed. The discussions for this rule and the previous rule can be applied to differential variances.

Imbalance in sample sizes, costs, and variances can all be assessed by these rules. On the whole, minor imbalances have minimal effects on costs and precision. Ordinarily initial considerations in study design can ignore these aspects and focus on the key aspects of estimation and variability.

2.11 USE THE RULE OF THREES TO CALCULATE 95% UPPER BOUNDS WHEN THERE HAVE BEEN NO EVENTS

Introduction

The rule of threes can be used to address the following type of question: "I am told by my physician that I need a serious operation and have been informed that there has not been a fatal outcome in the 20 operations carried out by the physician. Does this information give me an estimate of the potential postoperative mortality?" The answer is "yes!"

Rule of Thumb

Given no observed events in n trials, a 95% upper bound on the rate of occurrence is

$$\frac{3}{n}. \tag{2.40}$$

Illustration

Given no observed events in 20 trials a 95% upper bound on the rate of occurrence is $3/20 = 0.15$. Hence, with no fatalities in 20 operations the rate could still be as high as 0.15 or 15%.

Basis of the Rule

Formally, assume Y is Poisson (θ) using n samples. The Poisson has the useful property that the sum of independent Poisson variables is also Poisson. Hence in this case, $Y_1 + Y_2 + ... + Y_n$ is Poisson $(n\theta)$ and the question of at least one Y_i not equal to zero is the probability that the sum, $\sum Y_i$, is greater than zero. Specify this probability to be, say, 0.95 so that

$$P\left(\sum Y_i = 0\right) = e^{-n\theta} = 0.05. \tag{2.41}$$

Taking logarithms, produces

$$n\theta = -\ln(0.05) = 2.996 \sim 3 \tag{2.42}$$

Solving for θ,

$$\theta = \frac{3}{n}. \tag{2.43}$$

This is one version of the "rule of threes."

Discussion and Extensions

The equation, $n\theta = 3$, was solved for θ. It could have been solved for n as well. To illustrate this approach, consider the following question: "The concentration of Cryptosporidium in a water source is θ per liter. How many liters must I take to make sure that I have at least one organism?" The answer is, "Take $n = 3/\theta$ liters to be 95% certain that there is at least one organism in the sample."

Louis (1981) derived the rule of threes under the binomial model. He also points out that this value is the 95% Bayesian prediction interval using a uniform prior. For an interesting discussion see Hanley and Lippman-Hand (1983). For other applications see Fisher and van Belle (1993).

The key to the use of this result is that the number of trials without observed adverse events is known. The results have wide applicability. A similar rule can be derived for situations where one or more events are observed, but it is not as interesting as this situation.

2.12   SAMPLE SIZE CALCULATIONS SHOULD BE BASED ON THE WAY THE DATA WILL BE ANALYZED

There are obviously many ways to do sample size calculations, although the simple ones discussed in this chapter predominate. As a first approximation, calculate sample sizes for pairwise comparisons. While there are formulae for more complicated situations, they require specific alternative hypotheses which may not be acceptable or meaningful to the researcher.

Rule of Thumb

Ordinarily the sample size calculation should be based on the statistics used in the analysis of the data.

Illustration

If a sample size calculation is based on the normal model—that is, based on continuous data—then the data should not be analyzed by dichotomizing the observations and doing a binomial test.

Basis of the Rule

One of the key ingredients in sample size calculations is power which is associated with a particular statistical procedure. Carrying out some other statistical procedure during the analysis may alter the anticipated power. In addition, the estimated treatment effect may no longer be meaningful in the scale of the new statistic.

Discussion and Extensions

This rule is often more honored in the breach. For example, sample size calculations may be based on the two sample t-test of two treatments in a four-treatment completely randomized design, but an analysis of variance is actually carried out. This is probably the least offensive violation of the rule. The illustration above represents a more flagrant violation.

One situation where breaking the rule may make sense is where a more involved analysis will presumably increase the precision of the estimate and thus increase the power. This is a frequent strategy in grant applications where sample size calculations are required on short notice and the complexity of the data cannot be fully described at the time of application. For further discussion of this point see Friedman et al. (1998, page 99).

This chapter has shown the versatility of some simple formulae for calculating sample sizes. Equation (2.3) is the basis for many of the calculations. Equation (2.2) permits sample size calculations where the variances and the sample sizes are not equal. This approach needs only to be used when there are marked inequalities. For planning purposes it is best to start with the stronger assumption of equal sample sizes and equal variances.

3

Covariation

One of the key statistical tasks is the investigation of relationships among variables. Common terms used to describe them include agreement, concordance, contingency, correlation, association, and regression. In this chapter *covariation* will be used as the generic term indicating some kind of linkage between events or variables. Almost equivalent is the term *association*, but it tends to conjure up images of categorical data. Covariation is more neutral and also has the advantage of suggesting a statistical-scientific framework.

The topic of covariation is immense; a whole book could be written. Books on this topic typically fall into two categories: those dealing with discrete variables, for example Agresti (1990), and those dealing with continuous variables, for example Kleinbaum et al. (1998). Additionally, specific subject areas, such as epidemiology and environmental studies, tends to have their own specialized books.

The rules suggested here can be multiplied manyfold. The aim of this chapter is to discuss assumptions and some straightforward implications of ways of assessing covariation. One of the themes is that assessing covariation requires a good understanding of the research area and the purpose for which the measure of covariation is going to be assessed. Since the Pearson product-moment coefficient of correlation is the most widely used and abused, the chapter focuses on it.

It is unprofitable to attempt to devise a complete taxonomy of all the terms used to describe covariation. It will suffice to outline some broad categories as summarized in Figure 3. The first category deals with the source of the data: Are they based on experimental studies or observational studies? In a broad sense this deals with the way the data are generated. This will be important in terms of the appropriate inferences. A second classification describes the nature of the variables; a rough split is categorical versus continuous. This leaves ordinal variables and rank statistics in

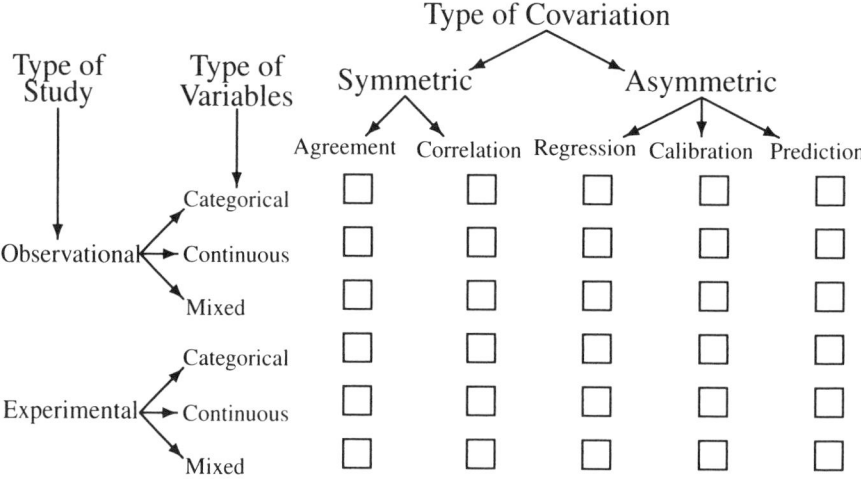

Fig. 3.1   Taxonomy of measures of covariation.

limbo territory somewhere between these two, but this will do at first blush. Things become complicated enough if one variable is categorical and the other is not. The third classification is symmetric versus asymmetric. Correlation is implicitly based on symmetric covariation, whereas regression is asymmetric. Within symmetric measures a distinction is made between agreement and correlation. Within asymmetric measures, distinguish among regression, calibration, and prediction. A measure of causal covariation will, by necessity, be asymmetric. Together these classifications lead to 30 subcategories, indicating that selecting appropriate measures of covariation is a nontrivial task. Specific scientific areas tend to have their own favorite measures. For example, the kappa statistic is very popular in the social sciences. Some measures can be used in several of these classifications. A key point is that some measures are inappropriate—or at least, less than optimal—for some situations, as will be indicated below.

This classification should be used as a rough guide. One reason for leaving the cells empty is there may be several measures that fit a particular category. The classification is intended to be a reflection starter—not a finisher. It would be interesting to have a group of statistical consultants recommend the most desirable statistical measure for each of the cells and assess their agreement.

One of the hardest statistical tasks is to move from association to causation. Many scientific questions involve causation: Does the increase in carbon dioxide in our atmosphere cause global warming? Does angioplasty reduce the risk of heart attacks? Similar questions are found daily in the press and on TV. These are difficult questions. To address them requires an interplay of statistics and the scientific discipline involved. Particularly difficult are analyses of associations that have a time dimension. One of the more famous is Doll's analysis of cigarette consumption in 1930 and lung cancer rates in 1950 for 11 countries (Doll, 1955). The implication of the analysis was that the association was causal. This analysis also involved other knotty statistical problems such as spurious correlation.

This chapter begins with a rule about sorting out the nature of the covariation as the first requirement for a valid analysis. Correlation is addressed next. Then it moves to discussing other types of covariation requiring other measures.

3.1 ASSESSING AND DESCRIBING COVARIATION

Introduction

As indicated above, there are many ways to describe or assess covariation. In a statistical context, covariation always involves variables, ideally random variables with known distributions.

Rule of Thumb

Before choosing a measure of covariation determine the source of the data (sampling scheme), the nature of the variables, and the symmetry status of the measure.

Illustration

A common measure of covariation is correlation. This is a symmetric measure; initially associated with continuous, normally distributed random variables. It has also been used for non-normal variables such as rank statistics and multinomial random variables. This measure is not appropriate for asymmetric situations, especially where the selection of one of the variables is not random. This is true whether the study is observational or experimental.

Basis of the Rule

Appropriateness of use: the wrong tools imply inappropriate assumptions, and may lead to incorrect or incomplete conclusions.

Discussion and Extensions

The taxonomy of measures of covariation is a good start for asking about the purpose of a study. This will result in clarification of study objectives. For example, if a new method is to be compared to a gold standard, the area is calibration. This situation is asymmetric for at least two reasons. First, the gold standard is the predictor variable, and the new method's values the dependent variable. Second, values of the gold standard may be chosen to cover a wide range to enable assessment of the usefulness of the new method. Correlation is inherently inappropriate in this situation.

When two equivalent scales are to be compared—for example, covariation associated with raters—then a symmetric measure is desirable. Two cases arise: Is the purpose of the assessment to indicate association, or is association assumed and its strength is of interest? In the former case a correlation is appropriate (assuming continuous data), whereas in the latter case a measure of agreement is more appropriate. Also, in the former case the scale of measurement is not critical, whereas in the latter it is.

For multivariate normal data the natural measures of covariation start with the variance–covariance matrix. If symmetry is involved, then correlations are appropriate; if asymmetry is involved, then regression coefficients are preferred. The sampling scheme frequently dictates the appropriate measure.

There are many more measures of association for categorical data than for continuous data. One reason is that there is more diversity in categorical data. A good discussion can be found in Agresti (1990) and references therein. The plethora of measures of association, along with the common practice of computer packages to list a host of them, leads to the danger of cherry picking: calculating all possible measures of association for a 2×2 table and then picking the one with the smallest p-value! But this is a temptation to which no one is completely immune.

A very popular measure of agreement in contingency tables in the social sciences is kappa; see Agresti (1990), Section 10.5.1. The measure is intended to adjust for marginal agreement; however, it does not really do so. There are additional problems with this statistic: The variance is a large-sample variance, and it does not lend itself to embedding in a statistical model—as for example, the logarithm of the odds ratio and the Pearson product-moment correlation.

This first rule is very useful for consideration at the beginning of an investigation. The investigator will have to determine the kind of study that will be carried out, what kinds of variables will be collected, and whether the associations are intrinsically symmetric or asymmetric.

3.2 DON'T SUMMARIZE REGRESSION SAMPLING SCHEMES WITH CORRELATION

Introduction

It is not commonly recognized that the estimate of correlation in a regression model and analysis must be handled carefully. The issue revolves around the sampling that leads to the data. Two sampling schemes are common. In *bivariate* sampling, subjects (or objects) are selected randomly from the population and X and Y are measured. Thus, both variables are random. For example, select 100 people from a population and measure their height and weight. In *regression* sampling, the researcher chooses the values of X and randomly selects subjects with the corresponding values and then observes Y. For example, randomly select from a population subjects with heights 60, 62, 64, 66, or 68 inches and measure their weight.

Rule of Thumb

Do not summarize regression sampling schemes with correlations.

Illustration

Analysis of variance is intimately related to regression. Many analysis of variance packages present the correlation, or "proportion of variance explained." These correlations and percentages are related to the choice of the values of X (see below) and therefore have limited meaning. Since an analysis of variance is intrinsically asymmetric, a correlational type of measure should not be used.

Basis of the Rule

Let (X_i, Y_i) be n pairs of observations in a regression sampling scheme with the X_i values under the control of the investigator. The following notation is useful:

$$[x^2] = \Sigma(x_i - \bar{x})^2, \tag{3.1}$$

$$[xy] = \Sigma(x_i - \bar{x})(y_i - \bar{y}), \tag{3.2}$$

and

$$[y^2] = \Sigma(y_i - \bar{y})^2. \tag{3.3}$$

The following relationships are the basis for the analysis of variance of regression relationships and the basis for the "proportion of variance explained."

$$(n - 2)s_{y.x}^2 = [y^2] - \frac{[xy]^2}{[x^2]} \tag{3.4}$$

and

$$r^2 = \frac{[xy]^2}{[x^2][y^2]} = b^2 \frac{[x^2]}{[y^2]}. \tag{3.5}$$

The statistic r^2 is the usual Pearson product-moment correlation coefficient. Substituting the last part of this expression in the previous equation and rearranging terms produces:

$$r^2 = \frac{1}{1 + \dfrac{(n - 2)s_{y.x}^2}{b^2[x^2]}}. \tag{3.6}$$

The quantities $s_{y.x}^2$ and b^2 are characteristics of the system being studied, and they are estimates of parameters that cannot be changed . However, the quantities $n - 2$ and $[x^2]$ are under the control of the investigator. X is associated with the design matrix and $n - 2$ represents the sampling intensity. In theory then, the value of $[x^2]$ can be made as large (or small) as desired. If the value of $[x^2]$ is made very large, then r^2 will tend to be close to 1, and r can be close to +1 or -1! In summary, r is dependent on the choice of values of X.

Discussion and Extensions

Suppose there are two designs. In the first design, subjects are randomly selected from the population of interest so that both X and Y are random. Let $[x^2]_{random}$ be the sum of squares of the X variable in this cases. In the second design, the investigator chooses the values of X; let $[x^2]_{regression}$ be the sum of squares of the X variable in this case. Assuming that the values $s_{y.x}^2$ and b are the same in both situations (which will be approximately true), then it can be shown that

$$\frac{r_{regression}^2}{1 - r_{regression}^2} = \frac{r_{random}^2}{1 - r_{random}^2} \frac{[x^2]_{regression}}{[x^2]_{random}}. \tag{3.7}$$

The correlations are the quantities estimated in the regression situation and in the random situation. The formula indicates that only if the spacing of the X values mimics that of the population will the two correlations be the same. If the regression spacing is larger, then the corresponding estimate of the correlation will be larger. Another way of stating this is to note that the variance of the X's in the sample determines the size of the estimate of the correlation coefficient; note that for any set of values of X, $(n - 1)s_x^2 = [x^2]$. To illustrate, suppose that the values of X are chosen to produce twice the true variance, then if $r_{random}^2 = 0.5$ the value for $r_{regression}^2$ will be 0.63; a 26% increase in the "proportion of variance explained" over the true proportion explained by the predictor variable.

It is probably impossible to avoid the use of correlations in regression sampling schemes. The practice is embedded in applied statistics. However, the user should be aware of its limitations, and should be able to apply alternative measures such as regression coefficients.

3.3   DO NOT CORRELATE RATES OR RATIOS INDISCRIMINATELY

Introduction

The correlation of rates raises the potential of "spurious correlation." Specifically, let X, Y, Z be three variables that are mutually independent so that pairwise correlations are zero. Then the ratios X/Z and Y/Z will be correlated due to the common denominator.

Rule of Thumb

Do not correlate rates or ratios indiscriminately.

Illustration

Neyman (1952), quoted in Kronmal (1993), provided an example of correlation between storks and babies. Neyman generated three statistically independent random variables for 54 "counties": number of women, number of babies, and number of storks. He then divided the numbers of babies and storks by the number of women to get two common population rates, correlated them, and obtained a correlation of 0.63 ($p < 0.00001$).

Basis of the Rule

Ratio variables with at least one common term are correlated.

Discussion and Extensions

The issue was first noted by K. Pearson in 1897. A discussion can be found in the *Encyclopedia of Statistical Sciences* under the heading of spurious correlation. Further discussion can be found in Kronmal (1993) and Neville and Holder (1995). Kronmal (1993) shows that a partial correlation of these ratios conditional on the reciprocal of the population size removes the spurious part of the correlation.

Another approach to avoiding the problem of spurious correlation uses regression on the logarithm of the variables. A common denominator can then be made one of the predictor variables and the spurious aspect of the problem goes away. An analogous approach is to use an exponential or multipredictive model. Cochrane et al. (1978a, 1978b) correlated for 18 developed countries the number of physicians per capita population and infant mortality rates, finding a positive correlation. Sankrithi et al. (1991) reanalyzed these data using a multiplicative model and found no correlation suggesting that the association was spurious. This is not the end of the story; Young (2001) studied similar rates in Japanese prefectures and counties in the United States. He did not use the multiplicative model because the linear model "is the simplest model and therefore the best starting point." Thus he, again, found significant correlations.

Rates and ratios are ubiquitous. Their covariation will continue to be assessed. Often this practice will be perfectly reasonable. However, statistical practitioners should always have the potential of spurious correlation in the back of their minds and know how to handle the problem when it crops up.

3.4 DETERMINING SAMPLE SIZE TO ESTIMATE A CORRELATION

Introduction

When bivariate observations are sampled from a population (therefore both X and Y are random), there is a simple way to calculate the sample size needed to detect a specified correlation or to compare two correlations from two independent samples.

Rule of Thumb

To determine the appropriate sample size to estimate a population correlation ρ, use Equation (2.3), page (31) with

$$\Delta = \frac{1}{2}\ln\frac{1+\rho}{1-\rho} \tag{3.8}$$

or use Figure 2.1, page (34).

Illustration

Suppose a correlation of $\rho = 0.4$ is to be detected, what is the sample size needed? For $\rho = 0.4$ the value of $\Delta = \frac{1}{2}\ln\frac{1.4}{0.6} = 0.4236$ and the estimated sample size (with Type I error 0.05 and power 0.8) is

$$n = \frac{8}{(0.4236)^2} = 44.6 \approx 45. \tag{3.9}$$

This calculation assumes a one-sample situation, hence the numerator is 8.

Basis of the Rule

The basis of this rule is the Fisher z-transformation. Given a sample of size n drawn from a bivariate normal population with correlation ρ, R.A. Fisher showed that

$$z = \frac{1}{2}\ln\left(\frac{1+r}{1-r}\right) \sim N\left(\frac{1}{2}\ln\left(\frac{1+\rho}{1-\rho}\right), \frac{1}{\sqrt{n-3}}\right). \tag{3.10}$$

The null hypothesis is that $\rho = 0$ and the alternative hypothesis has value ρ. Equation 2.3 becomes,

$$n - 3 = \frac{8}{\left(\frac{1}{2}\ln\frac{1+\rho}{1-\rho}\right)^2}. \tag{3.11}$$

Equation (2.3) and Figure 2.1 can be used if the quantity 3 is ignored (3 could be added to the number calculated using the equation; in practice it won't make much difference unless small sample sizes are considered).

Discussion and Extensions

The detection of small population correlations requires large sample sizes. For example, using equation (3.11) to detect a correlation of, say, $\rho = 0.05$ requires 3198 observation pairs. These kinds of sample sizes ordinarily do not occur in experimental work but may be seen in population studies. For example, current work relating mortality to air pollution, due to particulate matter, uses very large populations. Correlations of the order of 0.05 or even smaller are being investigated seriously. In survey sampling such small correlations can also occur.

Using correlations to calculate sample sizes is a good first step. However, more detailed knowledge of the proposed investigation will lead to additional considerations that may affect substantially the sample sizes needed.

3.5   PAIRING DATA IS NOT ALWAYS GOOD

Introduction

There is an old rule of thumb in the design of experiments that says that pairing data is always a good thing in that it increases precision. This is not so.

Rule of Thumb

Pairing should be investigated carefully. Do not pair unless the correlation between the pairs is greater than 0.5.

Illustration

Researchers frequently assume that adjusting for baseline value in a repeated measures experiment increases the precision. This is not always the case and may lead to less precision as illustrated below.

Basis of the Rule

To see this, consider the variance of the differences of paired data:

$$\text{Var}(Y_{i1} - Y_{i2}) = \sigma_1^2 + \sigma_2^2 - 2\rho\sigma_1\sigma_2, \tag{3.12}$$

where σ_i^2 is the variance of Y_i and ρ is the correlation. For simplicity assume that the two variances are equal so that

$$\text{Var}(Y_{i1} - Y_{i2}) = 2\sigma^2(1 - \rho). \tag{3.13}$$

If $\rho = 0$, that is, the variables are uncorrelated, then the variance of the difference is just equal to the sum of the variances. This variance is reduced *only* when $\rho > 0$: that is, there is a positive correlation between the two variables. Even in this case, the correlation ρ must be greater than 0.5 in order to have the variance of the difference be smaller than the variance of Y_{i2}. Such a large correlation is uncommon enough that it should always be checked.

Discussion and Extensions

One situation where the correlation between paired data will tend to be negative is where the total of the two observations is constrained in some way. For example, if two animals are caged together and have a common source of food, then the larger amount eaten by one animal will be offset by a smaller amount for the other animal. This will induce a negative correlation. Meier et al. (1958) give another example from a clinical trial situation. It may be necessary to adjust for a baseline value in order to standardize the observations. But when this is the case, a preliminary analysis should be carried out of the reasons for the discrepancy in the baseline values.

The discussion in this section has not required normality. To pursue this further, consider the situation where the observations are binary, such as case and matched control in an epidemiological study. Another example, involves the comparison of two methods of analyzing samples for environmental pollutants where the samples are scored as positive or negative. Using the latter scenario, assume n samples are to be analyzed by the two methods. The results can be arranged in a 2×2 table:

		Results by Method 2		
		Positive	Negative	
Results	**Positive**	n_{11}	n_{12}	$n_{1.}$
by Method 1	**Negative**	n_{21}	n_{22}	$n_{2.}$
		$n_{.1}$	$n_{.2}$	n

Several hypotheses are of interest. The hypothesis whether the proportion of "Positive" scores is the same for the two methods is usually examined by the McNemar test.

$$X^2 = \frac{(n_{21} - n_{12})^2}{(n_{21} + n_{12})}. \tag{3.14}$$

The test statistic is approximately chi-square with one degree of freedom. The McNemar test examines the off-diagonal elements; if the results are significant then the methods are considered not equivalent. This is also a test for differential accuracy.

Beginning students in statistics frequently carry out a "standard" chi-square test on this table. Does this test make any sense? The answer is yes. The usual chi-square test is a test of the hypothesis that the variables are correlated. Replace the values n_{ij} by their corresponding proportions:

		Method 2		
		Present	Absent	
Method 1	**Present**	p_{11}	p_{12}	$p_{1.}$
	Absent	p_{21}	p_{22}	$p_{2.}$
		$p_{.1}$	$p_{.2}$	1

Then the estimate of the ϕ coefficient is defined to be

$$\hat{\phi} = \frac{p_{11} - p_{1.}p_{.1}}{\sqrt{p_{1.}(1 - p_{1.})p_{.1}(1 - p_{.1})}}. \tag{3.15}$$

The usual chi-square test is given by

$$X^2 = n(\hat{\phi}^2). \tag{3.16}$$

The $\hat{\phi}$ coefficient is precisely the Pearson product-moment coefficient applied to the table of proportions, that is, $\hat{\phi} = \hat{\rho}$. The value of $\hat{\phi}$ can be positive or negative. Hence, if the matching was effective then $\hat{\phi} > 0$ and there was an advantage to matching. If the correlation is near zero or is negative then the matching was ineffective. If the correlation is negative then matching is, in fact, harmful in the sense that the variance of the difference is larger than the sum of the variances. So the two chi-square tests tell two stories about the binary data, and the first story should be about the usefulness of matching. The student who did the usual chi-square test on these data was the most sensible in first testing whether the observations were correlated, but also the most incomplete by not following through with the McNemar test.

The $\hat{\phi}$ coefficient is positive or negative depending on the sign of $p_{11} - p_{1.}p_{.1}$. That is, whether the observed proportion positive was greater or smaller than the expected proportion of positives.

One small extension. Consider two independent samples. Combine the samples and calculate the overall median. Score observations greater than the median positive, and observations less than the median negative. Then, under the null hypothesis of no difference in the sample medians, the expected row and column proportions in the associated contingency table are $1/2$ and the ϕ statistic reduces to

$$\hat{\phi} = 4p_{11} - 1. \tag{3.17}$$

When deviations are taken from the median, more than 25% of pairs must have both values positive in order for the correlation to be positive.

Pairing observations has intuitive appeal. This section indicates that such appeal should not prevent the researcher from checking whether this is useful in the study at hand. If the data are paired, the basic observational unit is usually the pair. Discussions of sample size, observation, treatment effect and so on, should make clear whether the pair is being discussed or one of the members of the pair.

3.6 GO BEYOND CORRELATION IN DRAWING CONCLUSIONS

Introduction

Correlations have limited usefulness in scientific settings because of their symmetrical nature. A great deal of scientific work investigates causal mechanisms and correlations are inherently unsuitable for such work.

Frequently researchers will use the correlation coefficient as measure of goodness-of-fit, association or some other characterization of a bivariate relationship. Several aspects of bivariate relationships are not adequately addressed by correlation. Specifically, the correlation coefficient is independent of changes in scale and location. Hence, an exact one-one correspondence between two variables as expressed in the original scales cannot be assessed by correlation. Similarly, differences in importance between two variables cannot be picked up by correlation because it is a symmetrical measure.

Rule of Thumb

Do not use the coefficient of correlation as a measure of agreement when location and scale are relevant.

Illustration

A situation in which location (and scale) are important is where two experts take the same readings, for example, blood pressures, or bone densities, or assay the same soil samples. In these cases the correlation could be very high but one expert reads consistently higher than the other, or one is more variable than the other even though both tend to estimate the same level. Appropriate measures of agreement and data analyses would clearly identify this.

Basis of the Rule

The invariance of the correlation coefficient under changes in location and scale is well known. Hence effects attributable to location and differential variability will not be detected.

Discussion and Extensions

The invariance of the correlation coefficient under changes in location and scale is both a virtue and a drawback. In some cases this invariance is useful, for example, when correlating temperature and product yield, the result should be independent of whether the temperature is read in degrees Fahrenheit or degrees Centigrade. The key question is, What is the nature of the association to be investigated? The statistical measures and tests have to be appropriate for the purpose. Another drawback of correlation is that rotation of correlational patterns will result in a different value for the coefficient of correlation, see Loh (1987) for illustrations. This is well-known for the situation where one variable in the correlation analysis is replaced by its negative so that the pattern is flipped 180^0 and the correlation changes sign; what is less well-known is that smaller rotations affect the value of the correlation also.

R.A. Fisher (1958) had cautionary advice about the use of correlation:

> In experimental work proper its [correlation coefficient] position is much less central; it will be found useful in the exploratory stages of an inquiry, as when two factors which had been thought independent appear to be associated in their occurrence; but it is seldom, with controlled experimental conditions, that it is desired to express our conclusion in the form of a correlation coefficient.

What Fisher had in mind was that regression or other quantitative estimates of covariation would be very much more informative.

Agreement is a more stringent concept than correlation since scales of the measurements and the slopes are important. A paper by Bland and Altman (1995) provides a useful discussion of measuring agreement in the case of clinical medicine. They suggest that if two measures X and Y are to be compared, analyze $X - Y$ and $X + Y$. A scatterplot of these two variables will be very informative indicating potential systematic bias, changes in bias for different values of the responses, changes in variability, and the identification of outliers. The next section provides a formal approach to measuring agreement.

If there is one statistical dogma, it is that correlation does not imply causation. No doubt, this is true. In addition, the researcher should not stop at correlation in measuring covariation.

3.7 ASSESS AGREEMENT IN TERMS OF ACCURACY, SCALE DIFFERENTIAL, AND PRECISION

Introduction

In comparing two methods of measuring a quantity, two questions are usually asked. First, do the methods agree? Second, how good is the agreement? These two questions form the background for the discussion of agreement. The first question involves an assessment of accuracy, the second precision. To make this operational this section will follow Lin (1989). Let two measurements of the same quantity, represented by Y_1 and Y_2, be bivariate normal $(\mu_1, \mu_2, \sigma_1, \sigma_2, \rho)$ representing the means, variances, and correlation coefficient. The idea is to explore the differences between the paired observations, $Y_1 - Y_2$. The square of these differences incorporates both location and scale effects as can be seen in the following formula,

$$E(Y_1 - Y_2)^2 = (\mu_1 - \mu_2)^2 + (\sigma_1 - \sigma_2)^2 + 2(1 - \rho)\sigma_1\sigma_2. \qquad (3.18)$$

If the observations agreed perfectly then they would fall on a 45^0 line through the origin. The average of the square of the distances of the paired values from this 45^0 line can be shown to be $0.5E(Y_1 - Y_2)^2$ hence equation (3.18) is promising for assessing components of the equation:

Rewrite equation (3.18) as follows and interpret the terms as indicated under the components of the equation:

$$\underbrace{\frac{E(Y_1 - Y_2)^2}{2\sigma_1\sigma_2}}_{\text{deviance}} = \underbrace{\frac{(\mu_1 - \mu_2)^2}{2\sigma_1\sigma_2}}_{\text{bias}} + \underbrace{\frac{(\sigma_1 - \sigma_2)^2}{2\sigma_1\sigma_2}}_{\text{scale diff.}} + \underbrace{(1 - \rho)}_{\text{imprecision}}. \qquad (3.19)$$

The terms on the right-hand side represent differentials in location (bias), scale, and precision. If both variables are relocated or rescaled by the same amount, there is no change in the values of these terms. If not, the effects will be detected by equation (3.19) which provides an anatomy of the concept of agreement. Note that the deviance is an omnibus statistic which is the sum of three other omnibus statistics, and it's these three that the researcher is interested in.

Rule of Thumb

Assess agreement by addressing accuracy, scale differential, and precision. Accuracy can be thought of as the lack of bias.

Illustration

Start with five "observations" 2,4,6,8,10. Multiplying by 2 produces 4,8,12,16,20. Adding 5 results in 9,13,17,21,25. The first and the last set now differ in scale and location but still have perfect correlation. The parameters in equation (3.19) can be estimated by the corresponding statistics,

$$\frac{\sum(Y_1 - Y_2)^2}{(n-1)2s_1s_2} = \frac{n(\bar{Y}_1 - \bar{Y}_2)^2}{(n-1)2s_1s_2} + \frac{(s_1 - s_2)^2}{2s_1s_2} + (1-r). \tag{3.20}$$

The terms in n and $n-1$ are needed because the standard deviation is based on the divisor $n-1$. For the data,

$$4.03125 = 3.78125 + 0.25 + (1\text{-}1).$$

This suggests that in this case the bias is much more important than the difference in scale. This example also illustrates that perfect correlation does not address agreement adequately.

Basis of the Rule

The statistical model provides the basis of the rule.

Discussion and Extensions

The contributions to the sample deviance can be tested formally in analysis of variance fashion with partitioning of the sum of squares at the total level,

$$\sum(Y_1 - Y_2)^2 = n(\bar{Y}_1 - \bar{Y}_2)^2 + (\sqrt{[Y_1^2]} - \sqrt{[Y_2^2]})^2 + 2(1-r)\sqrt{[Y_1^2][Y_2^2]} \tag{3.21}$$

where $[Y_i^2] = \sum(Y_{ij} - \bar{Y}_i)^2$. This partitions the sample deviance into the three components of bias, scale differential, and imprecision.

This way of comparing measurements clearly indicates why the correlation coefficient is inadequate in assessing agreement because it ignores two other components of agreement: precision and scale differential.

The deviance is equal to zero when all the (non-negative) terms on the right-hand side are exactly zero. This occurs when the two means, μ_1 and μ_2, are equal, the standard deviations, σ_1 and σ_2, are equal, and the correlation, ρ, is equal to 1. When these three conditions are fulfilled, there is perfect agreement between the two measures. Therefore, there are three aspects to disagreement (or agreement) relating to location, variation, and correlation. The first two are related to the concept of *accuracy* and the last to *precision*. The smaller the values of these terms, the greater the accuracy and the precision. Note the last term in equation (3.19) has a maximum value of 2. This occurs when the correlation is -1, the extreme of linear imprecision—$Y_1 = -Y_2$. The first two terms can be arbitrarily large. Hence, bias may be more important than precision.

The following characterizations of the data are defined:

- Location shift (u) $= \dfrac{(\mu_1 - \mu_2)}{\sqrt{2\sigma_1\sigma_2}}$,

- Scale shift (v) $= \dfrac{(\sigma_1 - \sigma_2)}{\sqrt{2\sigma_1\sigma_2}}$.

Lin (1989) used these quantities to come up with measures of accuracy presented here in slightly modified form:

$$\text{Accuracy} = \frac{1}{1 + \dfrac{(\mu_1 - \mu_2)^2}{2\sigma_1\sigma_2} + \dfrac{(\sigma_1 - \sigma_2)^2}{2\sigma_1\sigma_2}}. \tag{3.22}$$

The *accuracy* is now a quantity that varies between 0 and 1, with a value of 1 when the location and scale differentials are 0. The precision is given by the correlation coefficient, ρ. A measure of *concordance* is then defined by

$$\text{Concordance} = \text{Accuracy} * \text{Precision}$$

It can be seen that the *concordance* is equal to 1 when there is no location differential, no scale differential, and perfect correlation between the two variables. But the concordance is an omnibus quantity (see Rule 1.3). Hence, interpretation is difficult when it is substantially different from 1. The rule to keep in mind is that when terms such as *agreement*, *concordance*, or *goodness-of-fit* are summarized by the correlation coefficient, then the data are allowed to tell only part of the story. Using some of the techniques discussed in this section will give the rest of the story.

It is sometimes useful to test whether the estimated differential effects of mean (μ_1 and μ_2) and scale (σ_1 and σ_2) are significant. The differential effect of means can be tested by the paired t-test. The differential effect of scale can be tested by comparing the observed slope (of either Y on X or X on Y) against the theoretical values of $\beta = 1$ since this is the value obtained when $\sigma_1 = \sigma_2$. That is, the slope of the regression line of Y_1 on Y_2 is only equal to 1 when the standard deviations are equal. The significance of the precision can be tested by comparing the sample correlation r with the population value $\rho = 0$. This last test is not very interesting since substantial precision is expected. A more useful approach sets up a confidence interval associated with the value r to indicate the estimated precision. This can be done using the Fisher z-transformation.

Basically, this approach has reformulated and reduced the bivariate normal model containing five parameters to a model with three parameters to address issues of interest in the agreement situation. This was done to sharpen the bivariate normal model in order to address the questions and, furthermore, allows formulation of issues when the normal model is not appropriate.

This discussion has not dealt with the issue of agreement with a "gold standard." As indicated in the introduction on page 55, this is a problem in calibration. The methodology described can be applied to this situation as well. Accuracy in this case can be assessed in terms of a positive or negative bias.

Agreement has been formulated implicitly in terms of a linear relationship between two variables since correlation measures only linear association. It would be possible to extend this kind of analysis to nonlinear relationships. But this formulation already indicates a rich and profitable way of exploring agreement; one way of moving beyond correlation.

Agreement has been shown to be a more stringent measure of covariation than correlation. Correlation has an honorable role in assessing precision, but cannot be expected to address bias and heterogeneous variability. Researchers should routinely think beyond correlation when the covariation is being investigated.

3.8 ASSESS TEST RELIABILITY BY MEANS OF AGREEMENT

Introduction

In the social sciences the concept of test reliability is very important; a key idea is how well subjects perform on tests and involves the idea of reproducibility. Some of the ideas in this section are based on Arnold (1992).

Rule of Thumb

Test reliability in the social sciences can be assessed by means of agreement.

Illustration

The classical measurement model is given by

$$Y_{ij} = T_i + \epsilon_{ij}, \tag{3.23}$$

where Y_{ij} is the observed score for subject i, T_i is the subject's true score, and ϵ_{ij} is the error for the subject on the jth administration of the test or the jth form of the test. This model assumes that the errors are uncorrelated with each other and with the true scores. The true scores are postulated to have a variance σ_T^2 and the errors within subjects have variance σ_e^2. Hence,

$$\mathrm{Var}(Y_{ij}) = \mathrm{Var}(T_i) + \mathrm{Var}(\varepsilon_{ij}), \tag{3.24}$$

or

$$\sigma_Y^2 = \sigma_T^2 + \sigma_e^2. \tag{3.25}$$

These are the assumptions underlying the classical measurement model. See, for example, Lord and Novick (1968).

Suppose that n subjects are tested twice on a test according to the following scheme:

Subject	Time 1	Time 2	True Value
1	Y_{11}	Y_{12}	T_1
2	Y_{21}	Y_{22}	T_2
.	.	.	.
.	.	.	.
.	.	.	.
n	Y_{n1}	Y_{n2}	T_n

Here, Y is the observed score for a subject and T is the true score. Now assume that for a particular subject the scores are normally distributed about the true score, and that the true scores are also normally distributed. This leads to the following model for pairs of observations on a subject:

$$(Y_{i1}, Y_{i2}) \sim N\left(\mu, \mu, \sigma_T^2 + \sigma_e^2, \sigma_T^2 + \sigma_e^2, \sqrt{\frac{\sigma_T^2}{\sigma_T^2 + \sigma_e^2}}\right). \tag{3.26}$$

The quantity $\dfrac{\sigma_T^2}{\sigma_T^2 + \sigma_e^2}$ is called the *reliability* in the social sciences. In terms of the concepts and definitions from the previous section, the correlation coefficient ρ is equal to the reliability. Thus, reliability and precision are linked in a simple way.

Basis of the Rule

The derivation above constitutes the basis of the rule. Note that the test–retest model assumes that the means are equal and that variances within subjects are homogeneous.

Discussion and Extensions

In the above context two questions can be asked: Does the test that is being used measure what it is supposed to measure? If so, how well does it measure the true quantity? These are questions of validity and reliability, identical to the concepts of accuracy and precision defined earlier in this chapter. Hence, the whole discussion on accuracy and precision can be translated to the social science domain of validity and reliability.

The concepts of bias and differential scale variability have not been considered extensively in the social sciences because it becomes very difficult to estimate these in simple ways. Arnold (1992) and van Belle and Arnold (1999) consider violations of the assumptions of complete accuracy and differential scale variability on estimates of validity and reliability.

This formulation of tests has assumed that all the items in the test estimate the same quantity. An entirely different set of questions deals with the analysis of items on a test assuming that items have their own characteristics, *Item response theory* examines this area. It assumes that items vary in difficulty and individuals vary in ability to respond to the items. These questions lead to an entirely different statistical formulation. See Embretson and Reise (2000) and Husted et al. (2000) for useful formulations and discussions. These authors also address the question of equivalence of tests by scaling items on the tests in terms of difficulty and then equating test items with the same difficulty. Logistic distributions form a natural family for modeling these test situations.

This rule illustrates that statistical methods have wide applicability and that terms in one research area may have the same statistical meanings as different terms in another area. It also illustrates that different questions can lead to very different statistical formulations.

3.9   THE RANGE OF THE PREDICTOR VARIABLE DETERMINES THE PRECISION OF REGRESSION

Introduction

One of the basic models for covariation is the simple linear regression model based on the bivariate normal distribution. Suppose the following model is used:

$$Y = \alpha + \beta X + \epsilon, \tag{3.27}$$

where Y is the dependent variable, X is the predictor variable, α is the intercept, β is the slope, and ϵ the error which is assumed to be $N(0, \sigma^2)$. The errors are assumed to be independent. In many situations the values of X are chosen by the investigator—for example, doses in a bioassay. Just as the choice of values of X affects the estimation of the correlation coefficient (Rule 3.2), so the choice of X values also affects the precision of the estimate of the slope.

Rule of Thumb

Suppose the investigator has n pairs of observations available, with a specified range of values for the values of X. To get the smallest standard error for the estimate of the slope, place $n/2$ observations at the smallest value of X and the other $n/2$ observations at the largest value of X.

Illustration

Suppose that the range of values of X (or the range of scientific interest) is from 10 to 20 for an application of the linear model above. Suppose that 30 observations can be taken. The rule places 15 observations at the low end of the range, and 15 observations at the high end.

Basis of the Rule

The formula for the variance of the estimate, b, of the slope β is

$$\text{Var}(b) = \frac{\sigma^2_{y \cdot x}}{[x^2]}, \tag{3.28}$$

where

$$[x^2] = \Sigma (x_i - \bar{x})^2 \tag{3.29}$$

with the subscript i running over the n observations. The variance, $\sigma^2_{y \cdot x}$, estimated by the residual from regression, $s^2_{y \cdot x}$, is a characteristic of the biological system and can only be estimated but does not change. However, $[x^2]$ may be subject to the control of the investigator in the sense that the values of X can be chosen. This will control the variance of b, the estimate of β. Let X_{max} and X_{min} be the largest and smallest values of X, respectively. The value of $[x^2]$ is maximized when half of the n observations are put at X_{max} and half at X_{min} (for simplicity assume that n is even). For these values,

$$[x^2] = \frac{n}{4}(X_{max} - X_{min})^2. \tag{3.30}$$

Hence, from the point of view of the variance of the estimate of β, put half of the observations at the lowest value and half at the highest value of X.

Discussion and Extensions

There is much to recommend the above design but there may be one fatal flaw: the relationship between Y and X may not be linear. A design is desirable that "protects" against this possibility by permitting a check on the linearity. One possible design is to put the values of X at three places. The best design in this case, in terms of minimum variance, distributes the observations equally among the lowest, middle and highest values of X. Let $\text{Var}(b)_1$ and $\text{Var}(b)_2$ be the variances of the first design ($n/2$ observations at two points) and the second design ($n/3$ observations at three points), respectively. Then,

$$\frac{\text{Var}(b)_1}{\text{Var}(b)_2} = 2/3, \tag{3.31}$$

and $\sqrt{2/3} = 0.82$. So the price of being able to detect the simplest kind of nonlinearity (a quadratic relationship) involves an 18% increase in the standard error of the estimate of the slope. Spacing the values of X at four points allows a test of a cubic relationship; and the reduction in the standard error, compared with the two-point design, can be shown to be 25%. These results suggest two rules of thumb:

1. Protection is never free.

2. Experimental design does allow some control over variability of estimates.

Statistical aspects of the design of bioassays, a particular case of regression, go back to the early part of the twentieth century. The book by Finney (1978) is a classic in this area. He deals extensively with the design of two, three and four point assays. Gelman (2000) argues against observations at intermediate points of the range of the predictor variable since there will be little power to detect nonlinearity. This suggests that unless the model is known to be nonlinear (from perhaps mechanistic considerations), there is little advantage in planning for intermediate observations.

This rule is a good illustration of making the most of experimental control when it is available. It also demonstrates that there may be trade-offs. These should then be consciously made as determined by the objectives of the investigation. This issue is explored further in the next rule.

3.10 IN MEASURING CHANGE, WIDTH IS MORE IMPORTANT THAN NUMBER OF OBSERVATIONS

Introduction

The linear regression model can also be applied to situations involving change within an individual. Although the model is simple, there is a useful rule of thumb that can be derived. For example, in Alzheimer's disease the change in cognitive function is of considerable interest. This is measured by neuropsychological tests, usually scheduled at fixed intervals. The issue is, How often should such tests be administered to get a good idea of cognitive change? What is the best mix of frequency and spacing?

Rule of Thumb

In assessing change, the spacing of the observations is much more important than the number of observations.

Illustration

Consider two possible strategies in studying cognitive decline in Alzheimer's patients: The first is to observe a subject at baseline and 18 months. The second is to observe the subject several times within a year (12 months) in order to get the same precision as that obtained by the first approach (this would be of interest in a clinical trial of a new drug for the treatment of the disease). The result is that the subject would have to be examined monthly in order to get the same precision of the estimate of change as the first design where the subject is measured at baseline and 18 months.

Basis of the Rule

In the context of linear change, the slope of the line represents the change within an individual. Suppose that there are m observations within an individual spaced at equal time intervals. The formula for $[x^2]$ is the key component in the variance or standard error of the change within a specific individual. Suppose that the total spacing width is d (for example, $d = 12$ months). For the m observations,

$$[x^2] = \frac{d^2}{4} \frac{m(m+1)}{2(m-1)}. \qquad (3.32)$$

In terms of precision, consider the square root:

$$\sqrt{[x^2]} = \frac{d}{2} \sqrt{\frac{m(m+1)}{2(m-1)}}. \qquad (3.33)$$

The precision increases proportionately with d but essentially only as the square root of m, the number of observations per individual. In the example above, for the design with two observations at 0 and 18 months, $d = 18$ and $m = 2$. For the second design, $d = 12$ and m must be calculated. Working in the variance scale,

$$\frac{18^2}{4} \frac{2(3)}{3(1)} = \frac{12^2}{4} \frac{m(m+1)}{2(m-1)}. \qquad (3.34)$$

Solving the quadratic equation for m produces $m = 11.3 = 12$. If subjects are only observed for a year, they would have to be tested monthly to get the same estimate of precision of change in cognitive status as observing them once at entry and once at 18 months. Monthly testing is out of the question for practical and theoretical reasons (such as a learning effect). The greatest increase in precision occurs when observations are added at the extremes. In summary, when possible, always increase the spacing when assessing change; the number of observations are of secondary importance. This effect was already noted by R.A. Fisher (1958). This rule of thumb is analogous to the sample size results for the Poisson distribution discussed in Rule 2.6, page (40).

There are valid reasons for taking additional observations between baseline and last time of observation, such as making sure that there are no unexpected changes in the endpoint. But repeated observations cannot be justified on the basis of increase in precision. In fact, the investigator may want to save testing time (which is usually quite expensive) and use the funds saved to strengthen relationships with the patients, with the objective of reducing the probability of dropout.

Another possible reason for increasing the number of within-subject observations is that the within-subject variance must be estimated from the data. But in the case of change within subjects the error term is usually pooled across subjects so that it is more useful to have relatively few repeated observations within a subject but many subjects. This will also provide a better estimate of between-subject variability. See van Belle et al. (1990) for a more detailed discussion of estimating change in cognitive status in Alzheimer's disease.

The estimate of variability of change within a person using the simple linear regression model can be written as

$$\text{Var}(b) = \frac{\sigma_{y \cdot x}^2}{[x^2]} = \frac{\sigma_y^2 (1 - \rho)}{(n-1)s_x^2}, \qquad (3.35)$$

where σ_y^2 is the between-subject variance, s_x^2 is the variability in the predictor variable under the control of the investigator, and ρ is the correlation of observations within subjects. Sample size calculations can then be made for estimating change within a subject or, by extension, comparison of average change between two groups of subjects—for example, cognitive decline in healthy older adults and those suffering from a dementing illness. Diggle et al. (1994) have an extensive discussion of this topic.

The rules in this chapter indicate again the opportunities that a researcher has to increase the precision of a study without increased effort— except for some careful quantitative thinking. Trade-offs should be made deliberately and consciously.

4

Epidemiology

Epidemiology is the study of the diseases in populations. In general, epidemiologists ask four types of questions:

1. What is the incidence of the disease?

2. What is the prevalence of the disease?

3. What is the duration of the disease?

4. What are risk factors and risks associated with the disease?

Incidence is defined as the number, or proportion, of new cases of a disease occurring in a population per unit time. Prevalence is the number, or proportion, of cases that exist in a population at a particular time. Incidence and duration can act independently; together, they determine prevalence. The four key questions lead to a host of subsidiary issues such as definition and detection of the disease, competing risks, survival patterns, and preclinical screening. Clinical trials, especially those involving survival as an endpoint, are also of great epidemiological interest. All these issues are studied in order to come to a basic understanding of the mechanism and control of diseases.

Most diseases affect a relatively small proportion of the population at any one time. This chapter makes the "rare disease" assumption. (In some of the discussion sections alternatives to this assumption will be pointed out.) One characteristic of epidemiological studies is the emphasis on binary outcomes such as case or control, exposed or unexposed, risk factor present or risk factor absent, and survival or death. The logistic distribution is a natural statistical model for these kinds of variables.

Survival analysis models consider the incidence of disease over time. One possible model for such data is the Cox regression model.

Key measures of susceptibility to disease are relative risk and the odds ratio (see below). Under the rare disease assumption these two quantities are nearly equal and many epidemiological texts interchange the two (see Rule 4.2). An outstanding reference for a likelihood based approach to epidemiology is the text by Clayton and Hills (1993). Another very good text is by Selvin (1996). Books by Lachin (2000) and Needham (2001) are very comprehensive and at a higher mathematical level than the other two.

4.1 START WITH THE POISSON TO MODEL INCIDENCE OR PREVALENCE

Introduction

The occurrence of a particular disease, as measured by incidence or prevalence, is a relatively rare event. There are at least two reasons for this. First, the majority of a population is reasonably healthy. Second, there are many diseases, so that a particular disease will be rare.

Rule of Thumb

Start with the Poisson distribution to model disease incidence or prevalence.

Illustration

In the state of Washington in 1999, 11,706 persons were hospitalized with a primary diagnosis of acute myocardial infarction. Of these, 933 died (epiTRENDS, 2001) resulting in a rate of 8.0%. To construct a confidence interval on this rate assume that the number of deaths follows a Poisson distribution. Then the standard error is estimated to be $\sqrt{933} = 30.5$ deaths. A 95% confidence interval on the number of deaths is then $933 \pm 1.96 \times 30.5$ or (873, 993). This produces a 95% confidence interval on the rate of 8.% of (7.5%, 8.5%).

Basis of the Rule

The Poisson distribution as discussed in Rule 2.6, page (40), is associated with rare events. The Poisson, as that rule indicated, has the property that the variance is equal to the mean.

Discussion and Extensions

The Poisson distribution is a reasonable model for incidence and prevalence .Even for sampling situations which are modeled by a binomial distribution, the Poisson is a reasonable approximation to the binomial.

Key assumptions for the validity of a Poisson model for disease incidence or prevalence is that the occurrences of disease are independent and constant over time. These conditions provide a useful basis for exploration if the Poisson model does not hold. For example, contagious diseases (resulting in clustering) or changes over time of the occurrence of a disease.

The equality of mean and variance of the Poisson distribution provides a simple means of testing the validity of the model. If the variance is larger than the mean, the phrase *extra-Poisson variation* is used. In this situation a lognormal model is frequently considered. See Rule 4.4, page (84), for an illustration of the application of the lognormal distribution.

The Poisson model can also be used when there are a number of events in a specified number of person years of time. For example, the number of work-related fatalities in some industry over a total length of observed times of all the workers. Similar calculations as used in the illustration can be carried out for obtaining confidence intervals for the number of fatalities per units of person-years.

The Poisson distribution presents a powerful tool to epidemiology. It is a one-parameter model that is surprisingly useful.

4.2 THE ODDS RATIO APPROXIMATES THE RELATIVE RISK ASSUMING THE DISEASE IS RARE

Introduction

Three sampling approaches provide a statistical basis for observational epidemiological studies. In cross-sectional sampling, a random sample is drawn from a population and subjects are simultaneously classified with respect to disease status (D vs. $\overline{D}$) and risk factor or exposure status (E vs. $\overline{E}$). This may be called a prevalence study. In prospective or cohort studies, subjects are selected and followed for a specified length of time and classified as to disease status at the end of the study. There are several variations on this theme. Disease rates could be compared with a standardized population, for example occurrence of disease in a cohort of workers compared with the rate in the general population. Alternatively, subjects in the cohort could be classified by some kind of exposure status at the beginning of the study and disease occurrence at the end of the study in the exposed group compared with that of the unexposed group. In retrospective studies, subjects with the disease (cases) and those without (controls) are selected and classified with respect to the history of exposure prior to disease onset. A primary interest in all these studies is comparative; for example, disease pattern in an exposed group compared with the pattern in an unexposed group. A very common numerical comparison is the relative risk: the ratio of the probability of

Table 4.1   Occurrences of SIDS as Frequencies[a]

	Disease Status		
Birth Order	**SIDS**	**No SIDS**	**Total**
>1	201	68,900	69,101
1	92	62,600	62,692
Total	293	131,500	131,793

[a]Data from Peterson et al. (1979).

disease given exposure to the probability of disease given no exposure—for example, the ratio of the probability of a smoker developing lung cancer to the probability of a nonsmoker developing lung cancer. In symbols:

$$R = \frac{P(D|E)}{P(D|\overline{E})}. \tag{4.1}$$

The odds ratio is also used to measure the increase in the odds of disease with exposure to a risk factor. It is defined by

$$O = \frac{P(D, E)P(\overline{D}, \overline{E})}{P(D, \overline{E})P(\overline{D}, E)}. \tag{4.2}$$

Rule of Thumb

Under the rare disease assumption the odds ratio approximates the relative risk.

Illustration

Peterson et al. (1979) studied the characteristics of the Sudden Infant Death Syndrome (SIDS): deaths of infants less than one year for which no cause could be assigned. Table 4.1 categorizes all births in King County, Washington, USA from families with more than one child, for the period 1969–1977 with respect to SIDS status (disease) and birth order (exposure). Calculating probabilities produces Table 4.2 (notice that there is some rounding error).

The risk of SIDS among later born infants is 0.00153/0.52431 = 0.002918. The risk of SIDS among first-born infants is 0.00070/0.47569 = 0.001472. Thus the relative risk, R, is

$$R = \frac{0.002918}{0.001472} = 1.98$$

and the odds ratio, O, is,

$$O = \frac{0.00153 \times 0.47499}{0.00070 \times 0.52279} = 1.99.$$

Thus the relative risk and the odds ratio are virtually identical in this situation.

Table 4.2 Occurrences of SIDS as Proportions[a]

	Disease Status		
Birth Order	**SIDS**	**No SIDS**	**Total**
>1	0.00153	0.52279	0.52431
1	0.00070	0.47499	0.47569
Total	0.00222	0.99778	1.0000

[a] Data from Peterson et al. (1979).

Table 4.3 Probabilities for a Cross-Sectional Sampling Scheme

	Disease Status		
Exposure Status	**Disease(D)**	**No Disease($\overline{D}$)**	**Total**
Exposed(E)	π_{11}	π_{12}	$\pi_{1\cdot}$
Not exposed($\overline{E}$)	π_{21}	π_{22}	$\pi_{2\cdot}$
Total	$\pi_{\cdot1}$	$\pi_{\cdot2}$	1

Basis of the Rule

The argument will be based on assuming a cross-sectional sampling scheme. Suppose that the population can be partitioned as in Table 4.3.

The relative risk is

$$R = \frac{\pi_{11}/(\pi_{11} + \pi_{12})}{\pi_{21}/(\pi_{21} + \pi_{22})}. \tag{4.3}$$

Under the rare disease assumption, $\pi_{11} << \pi_{12}$ and $\pi_{21} << \pi_{22}$ so that these terms can be neglected in the denominators, producing

$$R \approx \frac{\pi_{11}/(\pi_{12})}{\pi_{21}/(\pi_{22})} = \frac{\pi_{11}\pi_{22}}{\pi_{12}\pi_{21}} = O. \tag{4.4}$$

Discussion and Extensions

It can be shown that under the three sampling schemes mentioned above (cross-sectional, cohort, and case-control) the odds ratio can be estimated validly. However, the relative risk can only be estimated validly under the first two sampling schemes. Tables 4.4 and 4.5 make this clear by the fact that the marginal row or column probabilities are equal to 1 so the cell probabilities are conditional and the marginal probabilities cannot be estimated. This is one advantage of the odds ratio. Another

Table 4.4   Probabilities for a Cohort Sampling Scheme: Prospective Study

	Disease Status		
Exposure Status	**Disease(D)**	**No Disease($\overline{D}$)**	**Total**
Exposed(E)	$\pi_{11}/\pi_{1\cdot}$	$\pi_{12}/\pi_{1\cdot}$	1
Not exposed($\overline{E}$)	$\pi_{21}/\pi_{2\cdot}$	$\pi_{22}/\pi_{2\cdot}$	1

Table 4.5   Probabilities for a Case-Control Sampling Scheme: Retrospective Study

	Disease Status	
Exposure Status	**Disease(D)**	**No Disease($\overline{D}$)**
Exposed(E)	$\pi_{11}/\pi_{\cdot 1}$	$\pi_{12}/\pi_{\cdot 2}$
Not exposed($\overline{E}$)	$\pi_{21}/\pi_{\cdot 1}$	$\pi_{22}/\pi_{\cdot 2}$
Total	1	1

advantage is that logistic models use the logarithm of the odds ratio as a "natural" metric so a great deal of statistical modeling is done in the context of the odds ratio.

The odds ratio will always be farther away from 1 than the relative risk. Using the cross-sectional sampling scheme, produces

$$O = \frac{R(1 - \pi_0)}{1 - \pi_0 R}, \tag{4.5}$$

where $\pi_0 = \pi_{21}/(\pi_{21} + \pi_{22})$ is the probability of the event in the control or unexposed population, that is, it is the background rate. Figure 4.1 displays the relationship between O and R as a function of π_0 on a logarithmic scale with values of the odds ratio between 0.1 and 10.

Figure 4.1 has several interesting characteristics.

- The odds ratio is always farther from one than the relative risk. Hence, if the relative risk is greater than 1, then the odds ratio is greater than the relative risk. Conversely, if the relative risk is less than 1, then the odds ratio is smaller than the relative risk.

- For a background rate of $\pi_0 < 0.05$ there is little difference between the odds ratio and the relative risk.

- The relative risk can never exceed $1/\pi_0$.

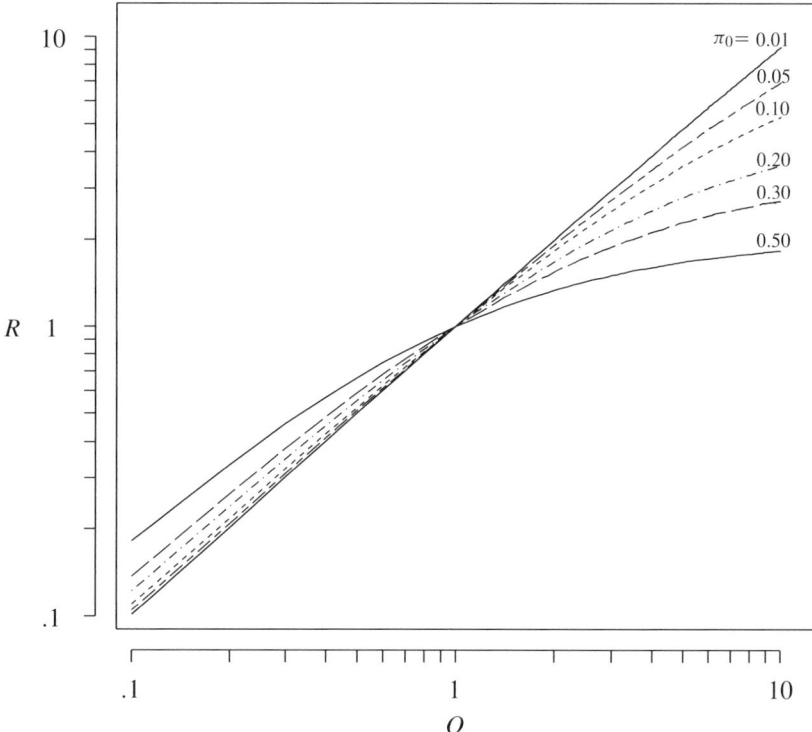

Fig. 4.1   Relationship between odds ratio O and relative risk R as a function of π_0, the background rate in the unexposed. Note that scale is logarithmic.

- The difference between the relative risk and the odds ratio is less pronounced when the relative risk and the odds ratio are less than 1.

- The "rare disease" assumption needs to hold in the unexposed population only for the relative risk and odds ratio to be approximately equal: the disease need not be rare in the exposed population (although that will tend to make for a very large relative risk).

- The rare disease assumption is a sufficient condition for the approximate equality of the odds ratio and the relative risk, it is not necessary. The odds ratio and relative risk are approximately equal near 1, regardless of the prevalence of the disease.

Table 4.6 lists the information that can be obtained from the three types of studies. Cohort studies are the most informative but also the most expensive, particularly when the disease is rare. While a cohort study can determine prevalence, it would take some time for the population to stabilize before prevalence can be determined.

Table 4.6 Risk Information in Three Types of Epidemiological Studies

Information	Type of Study		
	Cohort	Cross-Sectional	Case-Control
Incidence	yes	no	no
Prevalence	yes (uncommon)	yes	no
Relative risk	yes	yes	no
Odds ratio	yes	yes	yes

For this reason, cohort studies are commonly used to study incidence rather than prevalence. Cross-sectional studies are more economical than cohort studies. One drawback of a cross-sectional study is that it cannot determine the incidence of a disease. The case-control study is probably the cheapest to carry out. The most challenging aspect of a case-control study is the selection of appropriate controls.

Based on Lachin (2000), citing Bailar et al. (1984), approximately 46% of observational epidemiologic studies are cross-sectional, 29% are cohort studies, 6% are case-control studies, and the remainder other types such as case-studies. This classification does not include randomized, controlled clinical trials.

Many statistical packages implicitly assume an odds ratio formulation and thus can only be used for relative risk formulations under the rare disease assumption. The book by Clayton and Hills (1993) makes extensive use of the rare disease assumption. This assumption basically implies that a Poisson distribution can be used to model the epidemiological data. The assumption leads to surprisingly powerful epidemiological methods. The next section illustrates this approach.

The odds ratio is used most commonly in statistical analyses because of the extensive modeling possibilities. It is probably the most difficult to use in public communication. See Rule 4.12, page (100), for further discussion of the merits of several measures of epidemiological association.

4.3 THE NUMBER OF EVENTS IS CRUCIAL IN ESTIMATING SAMPLE SIZES

Introduction

Samples size questions come up often in epidemiological studies. As indicated in the discussion of Rule 4.2, the rare disease assumption permits application of the Poisson distribution and has implications for calculating sample sizes. This approach is illustrated by the following rule. A second, more common, approach uses a logarithmic model which will be discussed in Section 4.3.

Rule of Thumb

To detect a relative risk of R in a cohort study with relatively rare disease, the required number of exposed subjects, n (and the number of unexposed subjects, n) is given by

$$n = \frac{4}{\pi_0(\sqrt{R} - 1)^2},\tag{4.6}$$

where (using the probability structure of Table 4.4) $\pi_0 = \pi_{21}/\pi_2$. is the probability of disease in the unexposed population and R is the relative risk. Assume that $R > 1$. These sample sizes are associated with a two-sided Type I error of $\alpha = 0.05$ and power, $1 - \beta = 0.80$.

Illustration

In a cohort study it is estimated that 1/100 unexposed are at risk for a particular disease. What is the sample size required to detect a relative risk of $R = 3$ in a group exposed to a potential risk factor ? Using the formula produces

$$n = \frac{4}{0.01(\sqrt{3} - 1)^2} = 746.4 = 747.$$

So about 747 unexposed and 747 exposed are needed to detect a relative risk of 3.

Basis of the Rule

The rule is based on the Poisson model and the square root transformation. Under the alternative hypothesis, the numbers of unexposed with the disease and the number exposed with the disease are approximately Poisson $(n\pi_0)$ and Poisson $(n R\pi_0)$ respectively. The square root transformation stabilizes the variance at 0.5 (see Rule 2.6, page 40). The sample size formula then becomes

$$n = \frac{4}{(\sqrt{R\pi_0} - \sqrt{\pi_0})^2},\tag{4.7}$$

and the result follows.

Discussion and Extensions

The formula is reasonably accurate for disease rates among unexposed less than 0.2. For larger rates the usual formula for binomial sample sizes as discussed in Rule 2.8, page (43), should be used.

The formula indicates a dependence on the disease rate, π_0, in the unexposed population. If the rate can be doubled, the sample size can be reduced by a factor of two. This is sometimes possible when the observation period can be extended, for example, instead of observing event rates per year consider event rates per two or more years.

In the above discussion a distinction was made between the prevalence rate and the rate of occurrence of the disease in the unexposed. If the exposure is rare then these two rates will be very similar; for this reason, the distinction is not always maintained.

A further insight into the situation can be obtained by considering $n\pi_{21}$, the number of "events" in the unexposed group. Then the formula specifies the numbers of events that need to be observed in order to detect a relative risk of R. That is,

$$m = n\pi_0 = \frac{4}{(\sqrt{R} - 1)^2} \tag{4.8}$$

is the number of events in the unexposed, and Rm is the number of events in the exposed group needed to detect a relative risk of R. For example, to detect a relative risk of $R = 3$ requires $4/(\sqrt{3} - 1)^2 = 7.464$ events in the unexposed and $3 \times 7.464 = 22.4$ in the exposed group in order to have 80% power to detect a relative risk of 3. Rates of 1/100 in the unexposed and 3/100 in the exposed, require 747 unexposed and 747 exposed, as before. Hence it is the number of events that is crucial rather than the number of unexposed and exposed. This also suggests that it will be more efficient to divide the sample sizes for exposed and unexposed in such a way that the same number of events are observed in each group, rather than making the number of exposed equal to the number of unexposed. The precision of the estimate which is proportional to the reciprocal of the variances will be maximized when the variances are equal. This implies that the number of events must be equal in both groups.

The statement that the precision of estimates depends only on the number of events is only approximately true. For example, Hsieh (1989) has shown by simulation that increasing the number of subjects at risk while keeping the number of events constant will increase the power somewhat.

To summarize, the number of events is the crucial consideration in determining the precision of epidemiological studies. This has particular implications for studies with limited observation time. It is not the number of subjects that are included in the study but the number of events that are observed. Rule 4.8 contains additional discussion.

4.4 USING A LOGARITHMIC FORMULATION TO CALCULATE SAMPLE SIZE

Introduction

The more common approach to sample size calculation in epidemiological studies is based on the the logarithm of the relative risk, or the logarithm of the odds ratio. It gives comparable results over a large range of prevalences. Assume a cohort study with probability for disease π_0 in the unexposed and relative risk R for those exposed.

Rule of Thumb

The estimate of sample size per group in a cohort study, based on the logarithm of the relative risk, R, is

$$n = \frac{8(R+1)/R}{\pi_0 (\ln R)^2},$$ (4.9)

with power $1 - \beta = 0.80$, $\alpha = 0.05$, and a two-sided alternative.

Illustration

Consider again the example of the sample size required to detect a risk ratio of 3 when the background rate among unexposed is 1/100. Using the logarithmic approach, produces

$$n = \frac{8(3+1)/3}{\pi_0 (\ln 3)^2} = 883.8 \approx 884.$$

Basis of the Rule

The logarithm of the relative risk has approximate variance given by the following relationship, assuming equal sample sizes and the rare disease assumption,

$$n\mathrm{var}[\ln(R)] \approx \frac{1}{\pi_0} + \frac{1}{\pi_1}.$$ (4.10)

Using the relationship $\pi_1 = R\pi_0$ and substituting in the basic equation (2.3) with $(\mu_0 - \mu_1) = \ln(R)$ and $\sigma^2 = \frac{1}{\pi_0} + \frac{1}{R\pi_0}$ produces the result.

Discussion and Extensions

The derivation of the variance of $\ln(R)$ can be found in Lachin (2000) section 2.3.2. Equation (4.9) resembles equation (4.6); both have the factor π_0 in the denominator.

The required sample size for comparing two samples, based on the logarithm of the odds ratio, as opposed to the relative risk, is

$$n = \frac{8\mathrm{var}[\ln(O)]}{[\ln(O)]^2}.$$ (4.11)

The variance of this quantity is based on

$$n\mathrm{var}[\ln(O)] \approx \frac{1}{\pi_0} + \frac{1}{1-\pi_0} + \frac{1}{\pi_1} + \frac{1}{1-\pi_1}.$$ (4.12)

For the example above, estimate the variance by

$$n\mathrm{var}[\ln(O)] \approx \frac{1}{0.01} + \frac{1}{0.99} + \frac{1}{0.03} + \frac{1}{0.97} = 135.4,$$ (4.13)

and
$$\ln(O) = \ln \frac{0.03 \times 0.99}{0.01 \times 0.97} = 1.11902.$$
The sample size based on the logarithm of the odds ratio then becomes
$$n = \frac{8 \times 135.4}{(1.11903)^2} \approx 865.$$

The sample sizes produced by the square root formula, the logarithm of the risk ratio, and the logarithm of the odds ratio are 747, 884, and 865, respectively. These numbers would seem to differ substantially. However, these are only ballpark estimates of sample size. They give guidance as to rough order of magnitude.

The variance of $\ln(R)$ can be calculated from $1/0.01 + 1/0.03 = 133.3$. This is very close to the value for $n\text{var}[\ln(O)] = 135.4$. Similarly, $\ln(O) = 1.119$ and $\ln(R) = 1.099$. These values also validate the result that the odds ratio is always farther away from 1 when compared with the relative risk.

The problem could have been set up in terms of comparison of two binomials. Starting from the odds ratio formulation, derive
$$\pi_1 = \frac{O\pi_0}{1 + \pi_0(O - 1)} \qquad (4.14)$$
and then use appropriate formulae for binomial sample size calculations. For more extensive discussions of this approach to sample size estimates for epidemiological studies see Selvin (1996).

4.5 TAKE NO MORE THAN FOUR OR FIVE CONTROLS PER CASE

Introduction

The statistical structures of the cohort and case-control studies are similar as indicated in Tables 4.4 and 4.5 so similar sample size calculations can be applied. This section discusses the topic in terms of number of controls per case—an example of unequal sample size allocation introduced by Rule 2.9 page 45.

Rule of Thumb

In terms of variance, not much is gained beyond taking four or five controls per case.

Illustration

In a case-control study of Alzheimer's disease the number of cases was limited by the registry of the cases. Since the cases came from a large health maintenance organization, it was relatively easy to get additional controls (although not free). The recommendation was made to get three cases for each control as providing the "biggest bang for the buck." Equation (2.33) on page 46 indicates that for $k = 3$ the ratio of the standard errors is 0.81, compared with a maximum possible value of 0.71.

Basis of the Rule

The basis is equation (2.33) on page 46. This assumes equal variances in cases and controls.

Discussion and Extensions

The variance of the log of the relative risk (equation 4.10) and the log of the odds ratio (equation 4.12) differ only in the second and fourth terms. These are associated with the number of controls of the case-control structure. If the number of controls becomes very large, the variance of the $\ln(O)$ approaches that of $\ln(R)$. Suppose there are n_0 controls and n_1 cases then the variance of $\ln(O)$ is approximately,

$$\text{var}[\ln(O)] \approx \frac{1}{n_0 \pi_0} + \frac{1}{n_0(1 - \pi_0)} + \frac{1}{n_1 \pi_1} + \frac{1}{n_1(1 - \pi_1)}. \qquad (4.15)$$

Substituting $n_1 = kn_0$ leads to the result of equation (2.33)

This discussion links the precision of the cohort study with that of a case-control study and also indicates how the precision of the latter approaches that of the former. The variances of $\ln(R)$ and $\ln(O)$ depend only on the cell frequencies—not the marginal frequencies. This indicated, again, that the number of events is crucial to the precision of a study.

4.6 OBTAIN AT LEAST TEN SUBJECTS FOR EVERY VARIABLE INVESTIGATED

Introduction

A common question for statistical consultants is, "How many subjects (humans, animals) do I need as a function of the number of variables I want to study. When I asked several statisticians I got answers, but no source or justification for their answers." This section discusses some answers that have been given in the literature. It will become clear that the question can only be asked in specific contexts, which may generalize to other situations.

Rule of Thumb

In logistic regression situations about 10 events per variable are necessary in order to get reasonably stable estimates of the regression coefficients.

Illustration

In a five-year follow-up study of deaths following acute myocardial infarction about one-third of the patients are expected to die during the course of this study. How many subjects are needed if seven variables are considered as predictors of mortality?

By the rule of thumb about 70 events (deaths) are needed and therefore about 210 subjects should be enrolled. Note that this sample size deals only with stability of the estimates; it does *not* deal with power considerations.

Basis of the Rule

The rule is based on simulation studies by Peduzzi et al. (1996) involving data from the Department of Veterans Affairs Cooperative Study of Coronary Artery Surgery (see Peduzzi et al., 1985). This study followed 686 patients for a minimum of 10 years. At the time of the analysis, 252 deaths had occurred. Samples were generated with expected events per variable (EPV) of 2, 5, 10, 15, 20, and 25. The simulation assessed bias, precision and significance compared with the original sample. The authors state that for

> EPV values of 10 or greater, no major problems occurred. For EPV values less than 10, however, the regression coefficients were biased in both positive and negative directions; the large sample variance estimates from the logistic model both overestimated and underestimated the sample variance of the regressions coefficients; the 90% confidence limits about the estimated value did not have proper coverage; the Wald statistic was conservative under the null hypothesis; and paradoxical associations (significant in the wrong direction) were increased.

This result is quoted at length to indicate the many criteria that need to be considered when estimating the number of cases per variable.

Discussion and Extensions

The topic of the number of cases per variable has been discussed in the psychometric literature. A paper by Barcikowski and Stevens (1975) studied the stability of estimates of canonical correlations and came up with the requirement of 40 to 60 subjects per variable. Their evaluation dealt with continuous variables, so the results are somewhat sobering—for example, to get equivalent precision with dichotomous variables requires larger sample sizes. Thorndike (1978, page 184) gives more explicit formulae for the number of variables per case. He cites the following rule in the context of principal component analysis (PCA) :

> One informal guide (and perhaps lower limit) is that there should be 10 subjects for each variable. We should probably add 50 to this number to ensure sufficient sample size for small sets of variables. Thus our first rule of thumb is that $N \geq 10m + 50$.

Here m is used to denote the number of variables, Thorndike uses $p + c$ to indicate the total number of variables in the two sets of variables used in the canonical correlation. Thorndike also suggests a more stringent rule: $N \geq m^2 + 50$ resulting in still larger sample sizes.

The results in this section may be perceived as dismaying. They should lead to a healthy respect for adequate numbers of subjects if there are large numbers of variables. The emphasis was on stability of estimates; there was no discussion of power or the effect on Type I error rates due to the repeated testing of regression coefficients. This issue should also be considered in determining the number of subjects needed as related to the number of variables investigated.

4.7 BEGIN WITH THE EXPONENTIAL DISTRIBUTION TO MODEL TIME TO EVENT

Introduction

Rather than count the number of events in a certain time interval it is also possible to measure the time intervals between events, or the times of occurrence of events. The Poisson distribution is a strong model for counts. The exponential distribution is the basic model for survival times or times between events. Specifically, a random variable, W, has an exponential distribution with parameter λ if $P(W \leq t) = 1 - \exp(-\lambda t)$.

Rule of Thumb

The exponential distribution is the basic distribution for modeling survival times when the rate is constant over time.

Illustration

Lachin (2000) cites a study by Lewis et al. (1992) in which the survival of lupus patients was approximately exponential with parameter $\lambda = 0.3$. The mean and variance of the exponential distribution are $1/\lambda$ and $1/\lambda^2$, respectively. Thus, the average survival time, modeled by the exponential distribution, is $1/0.3 = 3.33$ years. The median survival time is that value of t such that $P(W \leq t) = 0.5$. Solving, $0.3t = -\ln 0.5$ and $t = 2.31$ years.

Basis of the Rule

The exponential distribution is the continuous analogue of the Poisson. Time intervals between events and the number of events are closely related as will be shown below.

Discussion and Extensions

Let W be the time until an event occurs. If W has the exponential distribution with parameter λ, that is, $P(W \leq t) = 1 - e^{-\lambda t}$, then

$$E(W) = \frac{1}{\lambda}, \tag{4.16}$$

$$Var(W) = \frac{1}{\lambda^2}. \tag{4.17}$$

There is a close connection between the Poisson and the exponential. Suppose, using the notation of Chapter 2, that Y is Poisson (θT), where θ is the rate per unit time and T is the length of time the process is observed. Then consider the Poisson event, $\{Y = 0$: No occurrences in the interval $T\}$, and the exponential event, $\{W > T$: survival is greater than $T\}$. Clearly, these two events are equivalent.

$$P(Y = 0) = e^{-\theta T} = P(W > T). \tag{4.18}$$

This leads to

$$P(W \le T) = F_W(T) = 1 - e^{-\lambda T}, \tag{4.19}$$

which is the cumulative distribution function of the exponential distribution. The quantity λ is called the *hazard rate*. A great deal of epidemiological modeling starts with this simple property. For example, assume an exponential survival time as above but assume that subjects are observed only for a time T and the number of events are observed in this interval. For explicitness assume that the event is time to death.

The exponential distribution is connected with a Poisson process. There is a specific relationship between the number of events, such as radioactive counts, that occur over time, and the time interval between these events. See Fraser (1976) , Kendall et al. (1987), Guttorp (1995) or any book on stochastic processes for a discussion of the relationship. The relationship between the Poisson and the exponential distribution leads to the link between incidence, prevalence and duration of disease. Assume that X, the number of new cases occurring per unit time, is Poisson with mean θ. Suppose the survival time, W, of these cases is exponential with parameter λ. Then the prevalence of the number of cases is Poisson θ/λ. Hence, if Y is the prevalence

$$E(Y) = \theta \left(\frac{1}{\lambda}\right) = E(X)E(W). \tag{4.20}$$

In words,

$$\text{Prevalence} = \text{Incidence} \times \text{Mean Duration of Disease.}$$

This assumes that survival times are independent of incidence. In other words, arrival and survival are independent. This Poisson model is an example of a queuing process. For example, customers arrive at a checkout stand in Poisson fashion (incidence), the service time is exponential (survival time), and the question is, how many checkout stands are occupied (prevalence)? The answer is given by the above relationship.

The example illustrates the wide applicability of the Poisson and exponential distributions. They form the basis for discussions of incidence, prevalence, and duration of disease.

The variance stabilizing transformation for the exponential distribution is logarithmic, as summarized in Table 4.7. This result is the basis for sample size calculations for survival times and comparisons of survival times. For comparison, the effect of a square root transformation is also presented in the table.

This section has shown that there is a close connection between the modeling of the number of events and the time intervals, or times, of the events. The connection is through the Poisson and exponential distributions. Assumptions made for one of the distributions have direct implications for the other distribution.

Table 4.7   Effect of Transformation on Mean and Variance of of an Exponential Random Variable

	X **Exponential**	Y $\ln X$	Y $\sqrt{X}$
Mean	$\dfrac{1}{\lambda}$	$-\ln\lambda$	$\sqrt{\dfrac{1}{\lambda}}$
Variance	$\left(\dfrac{1}{\lambda^2}\right)$	1	$\dfrac{1}{4\lambda}$

4.8   BEGIN WITH TWO EXPONENTIALS FOR COMPARING SURVIVAL TIMES

Introduction

Three rules of thumb are associated with applications of the exponential distribution; starting with the exponential distribution and then introducing two modifications: censoring and loss to follow-up. With these modifications a fairly rich epidemiological scenario can be handled. More detailed discussions and extensions can be found in the references cited at the beginning of the chapter.

Rule of Thumb

For comparing two populations with hazard functions λ_0 and λ_1, use sample sizes of

$$n = \frac{16}{[\ln(\lambda_1/\lambda_0)]^2} = \frac{16}{[\ln\lambda_0 - \ln\lambda_1]^2}, \tag{4.21}$$

where n is the number per group. The ratio λ_1/λ_0 is called the *hazard ratio*.

Illustration

Consider the lupus example again from page 89. Assume that the base hazard rate is 0.3. Suppose that a new treatment is expected to reduce this hazard to 0.18. This constitutes a 40% reduction in the hazard rate. Then using equation (4.21) the sample size per group is estimated to be,

$$n = \frac{16}{[\ln 0.3 - \ln 0.18]^2} = 61.3 = 62.$$

Thus, 62 patients are needed in the control group, and 62 patients in the experimental group.

Basis of the Rule

The rule is based on the logarithmic transformation of the exponential distribution discussed in Rule 4.7 on page 89. The variance of a single survival time is 1 and the result follows from equation (2.3) on page 31.

Discussion and Extensions

This model assumes that subjects are followed until they reach their endpoint. In the lupus example this is death. In practice, that is not the way clinical studies are carried out. Clinical trials cannot go on forever; there are funding and ethical constraints. Clinical trials involving times to an event are designed to stop after a fixed period of time. If patients are enrolled sequentially, there will be some who will be "on test" for a relatively short time until the trial ends—a variable time on test. One way to adjust for differential enrollment time in sample size calculations is to calculate the required number of events, the associated required observation time, and then the required number of patients. For example, if the number of required events in one of the groups is 100 with occurrence of the event anticipated as 1 per two years, then 200 patients need to be observed for two years, or 400 for a year. These are rough guidelines. More careful calculations can be done—for example, if subject age is an important consideration, over and above the length of time in the study. See Lachin (1998) and Newman (2001) for further details. This is an area where statisticians continue to do work.

Another source of reduced observation time is loss to follow-up or dropouts. Not all patients continue to the end of the trial. Investigating why some patients drop out is crucial. Often, the sicker patients drop out; or if a treatment has particularly serious side effects, patients in that group may drop out faster than those in other groups. A first adjustment to dropouts in sample size calculations is to increase the sample size in proportion to the dropout rate. If the dropout rate is expected to be 20%, so that 80% of subjects are expected to complete the study, then increase the calculated sample size per group, say n, to $n/0.80$. If the proportion of dropouts is expected to be much greater than 20%, the investigator will have to do careful review of the reasons for the dropouts and develop possible strategies for improving the retention rate. Again, see Lachin (1998) and Newman (2001) for further details.

Using the exponential distribution leads to very useful results which can be extended by making simple assumptions.

4.9   BE WARY OF SURROGATES

Introduction

In many studies it is inconvenient or too time-consuming to measure a clinical endpoint of interest and a substitute or surrogate is used—for example, CD4 counts are used in the study of HIV. In astronomy distance to stars is of great interest, but one cannot measure it directly; indirect measures such as relative brightness are used. A surrogate has been defined by Temple (1995) as follows:

> A surrogate endpoint of a clinical trial is a laboratory measurement or a physical sign used as a substitute for a clinically meaningful endpoint that measures directly how a patient feels, functions or survives. Changes induced by a therapy on a surrogate endpoint are expected to reflect changes in a clinically meaningful endpoint.

Rule of Thumb

Be wary of surrogates. Accept substitutes warily.

Illustration

The CD4 count has already been mentioned as a surrogate for measuring the course of HIV infection. Since AIDS is an immune disease, the more compromised the immune system, as measured by lower CD4 counts, the more severe the disease. The argument has a certain face validity. However, the CD4 count is not a perfect surrogate. Some of the drawbacks are that the CD4 count can be altered by infections such as colds, pneumonia, or influenza. The CD4 count is not changed immediately after infection with HIV, and exhibits variation within a day.

A second illustration is the use of the drugs encainide and flecainide to treat premature ventricular contractions (PVCs). It was known that PVCs were associated with increased risk of sudden death. Hence, the argument that suppression of PVCs was itself a desirable outcome. The drugs were approved on this basis. A subsequent randomized trial (Cardiac Arrythmia Suppression Trial, CAST) showed that there "was an excess of deaths due to arrhythmia and deaths due to shock after acute recurrent myocardial infarction in patients treated with encainide or flecainide" (Echt et al., 1991). This is one instance of a treatment with "face validity" of a surrogate that led to deleterious outcomes. See also Friedman et al. (1998).

Basis of the Rule

There are many instances where the surrogate did not behave as expected with respect to a clinical endpoint. That is, earlier studies using surrogates led to conclusions which subsequent studies (using hard endpoints) showed to be erroneous. Reliance on a surrogate may lead to wrong treatments. Caution is warranted.

Discussion and Extensions

A surrogate is tempting because it is easier to obtain than the outcome of interest. Frequently laboratory measurements, which are relatively easy to obtain, are used as surrogates. But, the clinical outcome (increased survival, lowered stroke rate, etc.) is of research and regulatory interest, not changes in laboratory measurements.

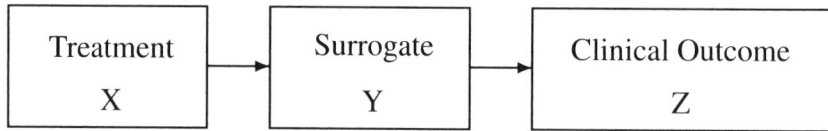

Fig. 4.2 Ideal surrogate where the surrogate is directly in the causal chain from treatment to clinical outcome. (From Zeger, 2000.)

Ideally a surrogate endpoint is directly in the causal chain between therapy and clinical outcome. This is the prescription of Prentice (Figure 4.2), as quoted by Zeger (2000), who pointed out that this is a very strict definition not likely to be met in any clinical situation. In the context of survival analysis, no surrogate will explain all the effects of the treatment. Even if it did, since there is measurement error, there will not be a perfect correlation. Zeger introduces the idea of a partial surrogate which, together with the clinical outcome, is correlated with an underlying latent variable. His advice with respect to surrogates is to focus, if at all possible, directly on the treatment and clinical outcome. He sees a useful role for surrogates in providing information when the treatment outcome cannot be measured directly, for example, when observations are censored.

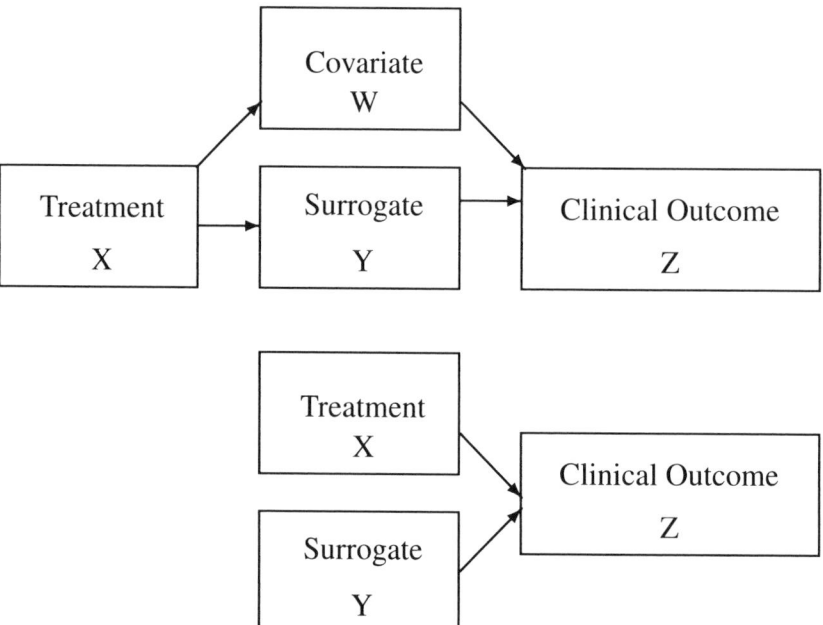

Fig. 4.3 Inadequate surrogates as suggested by N. R. Temkin (personal communication). In the first situation there is another variable, *W*, linked to the clinical outcome. In the second situation, treatment and surrogate are independent but both are linked to clinical outcome.

Inadequate surrogates are indicated the model in Figure 4.3. The first diagram indicates a surrogate affected by a treatment but not exclusively linked to the treatment. The second diagram indicates a surrogate associated with clinical outcome but not related to the treatment. In both of these instances it would have been a mistake to use the surrogate in place of the clinical outcome.

Surrogates may be reasonable for population level use, but not useful for clinical trial use. For example, lipid levels in a clinical trial are probably poorly associated with myocardial infarction. Hence, lipid level is a poor surrogate for treatment outcome. At the population level, however, lowering lipid levels may be a good assessment of treatment effect. See Pepe (2002) for an excellent discussion.

In observational studies surrogates of exposure can also occur. It may be difficult to separate surrogate exposures from the risk factor of interest; that is, surrogates that are in the causal chain between the risk factor of interest and the disease. If the surrogate is treated as a confounder, and adjusted for, the association between risk factor and disease may disappear. This would be a situation of over-adjustment.

A key arena for careful analysis of surrogates is where regulatory issues are involved. Here, the advantage of surrogates has to be balanced by the potential lack of clinical relevance. Regulatory agencies (such as the Food and Drug Administration in the United States) have to determine whether the time-saving surrogate used in a clinical trial is relevant to the disease outcome. See a useful discussion in Friedman et al. (1998). Similarly, in environmental studies there are huge debates about "indicator species." The spotted owl regulations in the United States have been the focal point for debates about environmental policy.

In another situation, occupational health and safety, the current debate in the United States about ergonomic standards is in many ways a debate about surrogates. The question is whether measurement of musculoskeletal movements characterize, or predict, clinical entities such as carpal tunnel syndrome. Statistical issues abound, ranging from measurement error to design of clinically relevant studies.

Surrogate endpoints will continue to be attractive and will continue to be used. Their use is not innocuous as the illustrations and discussion in this section have shown. For the user of surrogates the motto should be, *caveat emptor*.

4.10 PREVALENCE DOMINATES IN SCREENING RARE DISEASES

Introduction

One of the more important roles of public health is screening for disease. Programs such as screening for breast cancer, colon cancer, and prostate cancer come to mind. It is unfortunate that there is not a simple routine screening test for coronary artery disease even though many risk factors are known. The predictive value of screening program depends on three factors: the sensitivity and specificity of the screening procedure and the prevalence of the disease. A simple model for a screening procedure is given in Table 4.8. For convenience, denote the prevalence, $PREV$, of the disease by $\pi_0 (= \pi_{.1})$. The sensitivity and specificity of the test are defined by

Table 4.8   Model for Screening for Disease

| | **Disease Status** | | |
Screening Status	**Disease(D)**	**No Disease($\overline{D}$)**	**Total**
Test +	π_{11}	π_{12}	$\pi_{1.}$
Test −	π_{21}	π_{22}	$\pi_{2.}$
	$PREV = \pi_0$	$1 - PREV = 1 - \pi_0$	1

$$SENS = \frac{\pi_{11}}{\pi_0}, \tag{4.22}$$

and

$$SPEC = \frac{\pi_{22}}{1 - \pi_0}. \tag{4.23}$$

In words, the *sensitivity* is the probability that someone with the disease will be identified through the test, and the *specificity* is the probability that someone without the disease will be declared negative by the test. But what is frequently of interest is the probability of having the disease when the test is positive. This is called the positive predictive value (PPV) of the test:

$$PPV = \frac{\pi_{11}}{\pi_{1.}}. \tag{4.24}$$

Similarly, the probability of not having the disease when the test is negative is called the negative predictive value (NPV) of the test:

$$NPV = \frac{\pi_{22}}{\pi_{2.}}. \tag{4.25}$$

The quantities π_{21} and π_{12} are the probabilities of a false negative and false positive, respectively. It is easy to show that

$$PPV = \frac{\pi_{11}}{\pi_{1.}} = \frac{(SENS)(PREV)}{(SENS)(PREV) + (1 - SPEC)(1 - PREV)}. \tag{4.26}$$

Rule of Thumb

In rare diseases the prevalence dominates the predictive value of a positive test.

Example

Hoffrage et al. (2000) deal with screening for colorectal cancer. The prevalence of colorectal cancer is about 0.003. The hemoccult test has 50% sensitivity and 97% specificity. Then the predictive value of a positive test is 5%: that is, out of 100 people whose hemoccult test is positive, only five will have colorectal cancer.

Basis of the Rule

The basis for the rule can be seen most easily by expressing the equation in logits which maps the range of probabilities $[0,1]$ on to the whole real line $[-\infty, +\infty]$. For PPV it is defined by

$$\text{logit}(PPV) = \ln \frac{PPV}{1 - PPV}. \tag{4.27}$$

This can be related to the three components of PPV by using equation (4.26),

$$\frac{PPV}{1 - PPV} = \frac{SENS}{1 - SPEC} \frac{PREV}{1 - PREV}. \tag{4.28}$$

Taking logarithms on both sides produces

$$\text{logit}(PPV) = \ln \frac{SENS}{1 - SPEC} + \text{logit}(PREV). \tag{4.29}$$

For the example,

$$\text{logit}(PPV) = 2.813 - 5.806 = -2.993.$$

This produces $PPV = 0.048 \approx 5\%$.

Discussion and Extensions

Prevalence dominates: that is, for rare diseases $\text{logit}(PREV)$ is typically much less than $\ln(SENS)/(1 - SPEC)$ so that the logit of the predictive value of a positive test is less than 0, and the probability of disease, given a positive test, is less than 0.5. If the predictive value of a positive test is to be larger than 0.5, the logit must be positive. Under this condition,

$$\frac{SENS}{1 - SPEC} > \frac{1 - PREV}{PREV} \approx \frac{1}{PREV}. \tag{4.30}$$

This standard is very difficult to reach for rare diseases. To illustrate, suppose the sensitivity of a test is 100%. In order to have a greater that 50/50 chance that a person has the disease requires that

$$\frac{1}{1 - SPEC} > \frac{1}{PREV}, \tag{4.31}$$

or

$$SPEC > 1 - PREV. \tag{4.32}$$

For the data used in the illustration, in order to have a predictive value of a positive test greater than 50% (i.e. more likely than not), the specificity of the test must be $SPEC > 1 - PREV = 99.7\%$, which is very unlikely to be obtained in practice. Furthermore, the sensitivity of the homoccult test is much less than 100%.

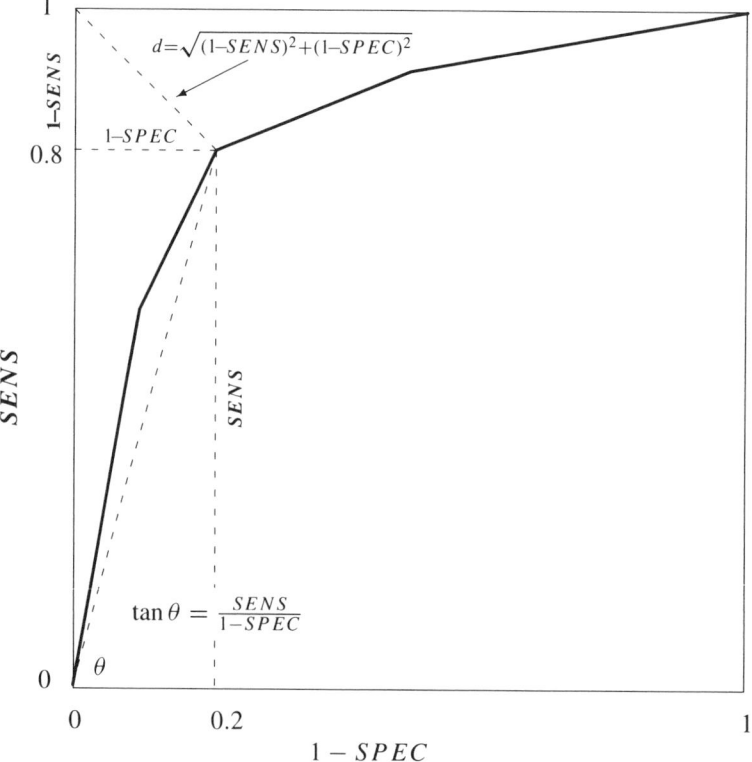

Fig. 4.4 ROC curve and some of its properties.

A specificity close to 100% is not unlikely for infectious diseases; the sensitivity tends to be be much less than 100%.

The quantity $SENS/(1 - SPEC)$ is a characteristic of the test. It is the likelihood ratio of a positive test, or positive diagnostic likelihood ratio (DLR); see Leisenring and Pepe (1999), and Boyko (1994). For an application of DLRs see Kerlikowske et al. (1996) or Sackett et al. (1997).

In some cases, the test is based on a continuous measurement, and cutoff values are defined to declare the test positive or negative. For example, in the case of blood pressure, high blood pressure is defined as a systolic blood pressure greater than 90 mm Hg. For every cutoff value there is a fixed value of the sensitivity and specificity. A plot of $SENS$ versus $(1 - SPEC)$ for all cutoff values is called a *receiver operating characteristic curve* or *ROC* curve. Figure 4.4 is an idealized version of such a curve. One question that arises is, What cutoff value should be chosen for the test? One idea is to choose the point on the curve closest to the upper left-hand corner since it represents the best possible situation: 100% sensitivity and specificity. Figure 4.4 shows that this is equivalent to minimizing $d^2 = (1 - SENS)^2 + (1 - SPEC)^2$. Another approach is to assign costs to false positives and false negatives and to minimize the expected cost; see Metz (1978) for an illustration. Specification of the rates may not be easy. In practice, a maximum false positive rate is maybe more easily specified (which has to be very small in screening) and this dictates the choice of cut-off point. There is a huge literature on ROC curves. Some topics are how to compare the curves of two different tests, how to estimate the curve from a small number of points, and the best criterion for choosing sensitivity and specificity. A very important question, reflecting clinical practice, is how to combine the information from several tests. When the PPV is low, a positive test needs to be confirmed with a more specific test. See Pepe (2003) and references for accessing the literature.

The effectiveness of a screening program depends on many factors such as reaching people at risk, patient acceptance, appropriate follow-up of positive screens, costs, transportation, and so on. The statistical characteristics prescribe the minimal requirements for success.

4.11   DO NOT DICHOTOMIZE UNLESS ABSOLUTELY NECESSARY

Introduction

As indicated in the introduction to this chapter epidemiologists like to dichotomize. This is natural because disease states and risk factors are usually thought of in a binary fashion. But dichotomization of continuous variables leads to loss of information.

Rule of Thumb

Do not dichotomize unless absolutely necessary.

Illustration

Suppose that high blood pressure is thought to be a risk factor for a disease. Subjects are dichotomized as having high blood pressure or not, and the odds ratio is calculated and tested for significance (Test A). An alternative approach tests the significance of the difference between the means blood pressures in the two groups by a two-sample t-test (Test B). Test B will be more powerful, that is, reject the null hypothesis of no difference when, in fact, it is false.

Basis of the Rule

The rule is based on the concept of asymptotic relative efficiency (ARE) of Test A compared with Test B. The ARE is the ratio of the sample sizes required for Test A and Test B to have the same power (assuming the same Type I error) when the sample sizes become very large. It has been shown that the two-sample sign test (Test A) above has ARE of $2/\pi = 0.64$. A study using 64 observations when analyzed with the t-test requires 100 observations when analyzed by the sign test. The dichotomization for the sign test is at the median. This comparison is for large sample sizes, for smaller sample sizes the efficiency is somewhat better.

Discussion and Extensions

The concept of ARE is discussed extensively in Hollander and Wolfe (1999) and Marascuilo and McSweeney (1977). Strictly speaking, the ARE mentioned in the example is associated with the two-sample median test which is slightly more powerful. The point of the rule is that, ordinarily, dichotomizing leads to inefficient use of information in the data. It should be pointed out that the ARE cited above compares the two-sample sign test with the normal model under normality assumptions: it's home territory, that is, where the t-test is expected to do well (but a similar result holds comparing the sign test with a rank test). There are models where the two-sample sign test is more efficient than the t-test, such as a double exponential. An extensive application to epidemiology can be found in Selvin (1996). He discusses a further drawback to dichotomization: What cut point will be used to dichotomize? He gives an example of an odds ratio changing substantially due to a change in cut point (Selvin 1996, page 99)

Screening tests may require dichotomization of a continuous endpoint because that is how the test will be used in practice. The sensitivity and the specificity of the dichotomized endpoints are then of interest. Dichotomizing may also be useful for for risk communication. In all these instances the choice of cut point is crucial and dependent on many kinds of factors.

Partitioning of continuous variables needs to be considered very carefully, ordinarily before the data are analyzed or based on (a priori) reasonable splits such as a median split.

4.12 SELECT AN ADDITIVE OR MULTIPLICATIVE MODEL ON THE BASIS OF MECHANISM OF ACTION

Introduction

The odds *ratio* and risk *ratio* (relative risk) indicate that odds and risk ratios are inherently multiplicative. In principle, there is no reason not to discuss *differences* in odds or rates. Logarithms of odds ratios are on an additive scale but clearly are multiplicative in the original scale. When should ratios be used rather than differences?

Rule of Thumb

Select an additive or multiplicative model according to the following order: theoretical justification, practical implication, and computer implementation.

Illustration

Rothman (1986) discusses some of the disadvantages of the multiplicative model. He gives the example of risk of lung cancer associated with cigarette smoking. If the risk factor is number of cigarettes smoked per day, then the multiplicative model implies that the change in risk going from 1 cigarette to 2 is smaller than the change in risk of going from 10 cigarettes to 11. This is a counterintuitive result casting some doubt on the usefulness of relative risk. Rothman also points out that the problem can be averted by dichotomizing the exposure variables.

Basis of the Rule

A basic requirement in a scientific study is to use a model that accurately reflects the "state of nature." Given that there is frequently ambiguity about this state the practical needs of the moment may dictate the choice of model. This, then, will also depend on the ability to implement the model—that is, find a computer routine that will carry out the analysis.

Discussion and Extensions

It is not possible to compare multiplicative and additive models directly since they are not nested. However, as pointed out by Clayton and Hills (1993), it is possible to embed these models in an *extended* model as follows:

$$\frac{(\text{Rate})^{\rho} - 1}{\rho} = \text{Intercept} + \text{Effect}(A) + \text{Effect}(B), \qquad (4.33)$$

where A and B are factors thought to be associated with the rate; for example, A could be Age, while B could be case or control status. This extended model has the property that as ρ approaches 1 the model is clearly additive and as ρ approaches zero the equation becomes

$$log(\text{Rate}) = \text{Intercept} + \text{Effect}(A) + \text{Effect}(B). \qquad (4.34)$$

thus becoming additive on the logarithmic scale and multiplicative in the rate scale. While this is an ingenious idea, it is difficult to implement computationally and has not found wide acceptance in computer packages. This approach was also discussed by Thomas (1981). The power of the ability to distinguish between the two classes of models depending on the value of ρ should be considered if the decision is made to go this route.

Public health implications of a set of risk factors would favor additive models since rates and changes in rates can be directly calculated in an additive model. But this is not an absolute necessity since slightly more complicated calculations in the multiplicative model will accomplish the same thing. The issue is not trivial. Clayton and Hills (1993) give an example (page 289) where usefulness of an intervention in a "high-risk" group depends on whether the model is additive or multiplicative.

Walter (2000) discusses the three most common methods of comparing risks: risk ratio, odds ratio , and risk difference. He points out that there is little numerical difference between the three ways of comparing risks over a wide range of outcome probabilities. He also points out that it will be difficult to discriminate between these methods on the basis of a statistical model. Given this overlap, other considerations will enter into the choice of risk expression such as ease of communication. For example, the risk difference is easier to understand than an odds ratio.

All the measures discussed in this chapter are omnibus quantities. The discussion in this section indicates that the selection of a particular expression of measures depends on statistical, epidemiological, and other considerations. Selection of a particular measure should be made before beginning the analysis. If the selection is data-driven, the reader should be advised of this.

5

Environmental Studies

5.1 THINK LOGNORMAL

This chapter discusses some statistical rules of thumb for environmental studies. There has been some concern that statisticians have not been adequately involved in such studies, see, for example, Millard (1987a). The concern arises particularly in the development and setting of policy (Guttorp, 2000). At its best, policy is the product of science, values, and politics. Science of this type has been called *mandated science* (Salter, 1988) and it is in this thorny arena that the statistician, as scientific colleague and collaborator, operates. The science is *mandated* because it involves the scientific basis for the policy; this must be obtained if it is lacking. Hence, regulatory agencies will fund studies that address policy-related science. For example, the Environmental Protection Agency of the United States government provides extramural funds for research under the STAR program: Science to achieve results. The acronym is a good illustration of mandated science. The phrase *sound science* also crops up in this arena. Usually with the implication that some of the science is unsound, that is, irrelevant or nitpicking. These are exciting areas but also frustrating because politicians look for immediate answers to problems of the day, which conflicts with scientific caution and appropriate hedging. Values may also conflict; and value clarification does not always lead to reconciliation. These considerations necessarily require statisticians to be outstanding methodologists, wise simplifiers, and great communicators.

An important strategy for beginning to reconcile conflicting values is to get agreement on the science. A very sensible and creative response has been to form jointly funded, but scientifically independent, organizations that sponsor research and provide a sound scientific basis and review of the evidence for the policy setters and those "sat upon." There are several nongovernmental organizations (NGOs) funded by groups with potentially different values and aims. One example is The Health Effects Institute (HEI) in Boston funded by the U.S. Environmental Protection Agency and a consortium of automobile and diesel manufacturers. This institute funds and reviews research investigating the health effects of automobile air pollution. Another example is the Mickey Leland National Urban Air Toxics Research Center (NUATRC) funded by the U.S. Environmental Protection Agency and a variety of chemical and oil producers. This center funds and reviews studies of health effects of air toxics. Statisticians play a prominent role in the design, review, and funding of environmental studies sponsored by these institutions. Their research products have been well-accepted by all sponsors—evidence of a successful integration of science and values. Regulators and policy setters then integrate these results into policy. The U.S. National Research Council is another independent source review; it does not sponsor research, but it can suggest research needs.

Environmental studies cover a huge area ranging from ecology to epidemiology to toxicology and risk assessment. In each of these areas there are outstanding statistical groups modeling and analyzing data.

5.2 BEGIN WITH THE LOGNORMAL DISTRIBUTION IN ENVIRONMENTAL STUDIES

Introduction

So far, extensive use has been made of the normal and Poisson statistical models. The normal distribution was the basis for most of the sample size calculations and regression models. The Poisson (and exponential) distributions were the basis for the previous chapter on epidemiology— as result of the 'rare disease' assumption and counting processes. The lognormal distribution was also used. In environmental studies the lognormal is a key distribution and this chapter starts with it. One reason is that environmental data tend to be skewed to the right, bounded on the left by zero, and lower and upper bounds are of special scientific and regulatory interest. Other approaches to dealing with skewed data will also be discussed.

Rule of Thumb

Think lognormal for measurement data in environmental studies.

Illustration

Blood pesticide levels in children living near orchards tend to be distributed lognormally. The levels tend to be low and bounded on the left by zero, with the added complications that levels of scientific interest are near the detection level and measurements are quite expensive.

Table 5.1 Characterizations of the Lognormal Distribution and Linkage to the Normal Distribution

	Lognormal Scale (X)	Original Scale (Y)
Mean	μ_x	$\mu_y = e^{\mu_x + \sigma_x^2/2}$
Median	$\tilde{\mu}_x = \mu_x$	$\tilde{\mu}_y = e^{\mu_x}$
Mean/Median	$\mu_x/\tilde{\mu}_x = 1$	$\mu_y/\tilde{\mu}_y = e^{\sigma_x^2/2}$
Variance	σ_x^2	$\sigma_y^2 = [e^{\sigma_x^2} - 1][e^{2\mu_x + \sigma_x^2}]$
CV^2	$CV_x^2 = \sigma_x^2/(\mu_x)^2$	$CV_y^2 = e^{\sigma_x^2} - 1$

Basis of the Rule

Ott (1995a) provides the basis for this rule. "A concentration undergoing a series of independent random dilutions tends to be lognormally distributed." Ott lists three conditions for a random process to form a lognormal process. Since many of these processes are associated with time, the conditions will be stated in terms of time.

1. The value of the outcome variable at a specified time can be expressed as a linear proportion of its value at a previous time.

2. Successive linear proportions are statistically independent.

3. The process has been going on for a long time.

Discussion and Extensions

Air and water concentrations in environmental studies are the result of independent dilution processes, which is equivalent to multiplying independent random variables rather than adding them. The lognormal distribution can be thought of as the central limit theorem distribution for these kinds of environmental variables. An extensive discussion can be found in Ott (1995a), Chapter 9.

The lognormal distribution can be described as follows: Let $X(= \ln Y) \sim N(\mu_x, \sigma_x^2)$, then $Y = e^X$ is lognormal. Table 5.1 lists some properties of the two distributions. Several useful rules can be derived from this table.

- The ratio of the mean to the median in the Y scale provides a crude test of lognormality since the log of this ratio must be approximately the variance in the logarithmic scale:

$$\sigma_x^2 = 2\ln\left(\frac{\mu_y}{\tilde{\mu}_y}\right). \tag{5.1}$$

- For a small coefficient of variation in the Y scale, it is approximately true that

$$\sigma_x \approx CV_y. \tag{5.2}$$

- For very large values of CV_y^2 (say, larger than 10),

$$\sigma_x^2 \approx \ln(CV_y^2). \tag{5.3}$$

- There is an interesting relationship between the mean, median, and coefficient of variation in the Y scale:

$$CV_y^2 = \left(\frac{\mu_y}{\widetilde{\mu}_y}\right)^2 - 1. \tag{5.4}$$

This relationship can be used as a crude check on the suitability of the log transformation since all three quantities can be estimated from the data.

- It has been pointed out (see Zhou and Gao, 1997, for some references) that a confidence interval for μ_x does not produce a confidence interval for μ_y by exponentiating the bounds on μ_x but produces a confidence interval for the median of the lognormal distribution, e^{μ_x}. A confidence interval for μ_y, needs to incorporate an interval estimate for σ_x^2. An estimate proposed by Cox (in Land 1972) for $\mu_x + \dfrac{\sigma_x^2}{2}$ is

$$\bar{X} + \frac{s_x^2}{2} \pm z_{1-\alpha/2}\sqrt{\frac{s^2}{n} + \frac{s^4}{2(n-1)}}, \tag{5.5}$$

where $\bar{X}, s_x^2$, and n are the mean, variance, and sample size of the observations in the log scale. The confidence interval for μ_y is then obtained by exponentiating. Zhou and Gao (1997) show that this estimator has reasonably good properties. They also recommend consideration of a bootstrap estimate of μ_y, which has superior properties for small sample sizes.

The lognormal distribution is a very versatile distribution and the natural one to consider when observations are bounded by zero. When a variable is bounded (below or above) the variance usually is related to the mean. From an inference point of view this is undesirable because estimate and precision of the estimate are usually considered separately, involving somewhat different statistical processes.

5.3 DIFFERENCES ARE MORE SYMMETRICAL

Introduction

Skewness of observations is a constant concern in environmental data because of the non-negative nature of most observational data. The lognormal distribution is a good example of this. Transformations are one way to introduce more symmetry into the distributions. Another approach deals with differences. In many situations, differences are of primary interest—for example, differences in methods of analysis of independent samples, emission of vehicles before and after installation of control equipment, and differences in FEV_1 (Forced Expiratory Volume after 1 second) between asthmatic persons and matched control subjects.

Rule of Thumb

Differences tend to be more symmetrical.

Illustration

There are two ways to measure fine particles in the air: One uses a filter technique, the other is based on light scattering using a nephelometer. A series of air samples are measured by both methods. One aspect of the analysis is whether one method systematically over- or under-estimates the mean level. While the original distributions of the samples may be quite skewed, the differences will tend to be much more symmetrically distributed. The usual normal-based tests can be applied to the differences.

Basis of the Rule

In most practical situations the difference will involve a location shift of the distribution, possibly with correlation of the original observations but with variances that are similar. In this case the distribution of the differences will be symmetrically distributed. If one of the variables in the difference dominates—that is, its variance and means are much greater than the other in the pair—it is clear that the skewness of the difference will be very similar to that of the dominant partner. This situation is usually not of practical interest.

Discussion and Extensions

The most common measure of skewness is the third moment about the mean normalized by a measure of spread. That is, for a random variable Y (assuming the third moment exists) the skewness γ_1 is defined by

$$\gamma_1 = \frac{E(Y - \mu)^3}{\sigma^3}. \tag{5.6}$$

A simpler measure of skewness that does not involve the third moment is

$$\frac{\text{mean} - \text{median}}{\sigma}. \tag{5.7}$$

One nice feature of this definition is that,

$$-1 \leq \frac{\text{mean} - \text{median}}{\sigma} \leq 1, \tag{5.8}$$

so that the mean and median never differ by more than one standard deviation. The inequality is sharp; that is, the bounds can be reached as shown by the simple example of a Bernoulli random variable, Y, with values of 0 and 1. The median is 0 when $P(Y = 0) = 0.5 + \epsilon$ where $\epsilon > 0$. In this situation,

$$\frac{\text{mean} - \text{median}}{\sigma} = \sqrt{\frac{0.5 - \epsilon}{0.5 + \epsilon}}. \tag{5.9}$$

This quantity can be made arbitrarily close to 1 so that the median is one standard deviation from the mean. This represents an extreme situation unlikely to be found in practice.

Variance stabilizing transformations frequently have the effect of making the transformed variable more symmetric. This is clearly the case for lognormally distributed data. This is also the case for the square root transformation of a Poisson random variable; see Fraser (1976).

If there is concern about asymmetry in a variable, there should be less concern when changes in the variable are considered. This is particularly true if the changes are with a location shift. In the case of bounded variables there will be a change in scale associated with a change in location. However, the change in scale is apt to be minor so that symmetric differences are still reasonable.

5.4 BEWARE OF PSEUDOREPLICATION

Introduction

The concept of independent samples is quite commonsensical—until faced with specific instances. For example, take a ventilation system with 100 duct openings. A ventilation engineer wants to test the efficiency of configurations of sizes of ducts. A particular arrangement is set up and the system is run. What constitutes independent samples? Separate runs on separate days? Is one sample the set of all air velocities at the duct openings?

In ecological statistics a similar problem arises—for example, sampling plankton blooms? What constitutes independent samples? In these two examples, independence has to be assessed in both time and space.

The term *pseudoreplication* is in the title of a paper by Hurlbert (1984). It is defined by lack of statistical independence. In the context of analysis of variance, Hurlbert defines pseudoreplication as the assessment of a treatment effect with "an error term inappropriate to the hypothesis being considered." Thus this problem is not unique to environmental studies. But it is particularly relevant here because many sampling studies involve measurements contiguous in time and/or space.

Rule of Thumb

Beware of pseudoreplication.

Illustration

The simplest example: Measuring hourly ozone levels for 24 hours in one day is not the same as measuring ozone levels one hour per day for 24 days. Or measuring one-hour ozone levels at 24 locations. All three sets contain 24 observations. It is the sampling procedure that determines the populations to which the inferences are to be made. Pseudoreplication occurs when the 24 samples in one day are treated as if they were representative of 24 daily observations. This illustrates Hurlbert's criterion about the variability having to be appropriate for the inferences to be made.

Basis of the Rule

In order to make valid inferences to a population, appropriate elements of the population must be selected with known probabilities. Pseudoreplication is a more important issue in interval estimation than point estimation.

Discussion and Extensions

Inadequate replication in space usually involves sampling too narrowly geographically so that the full variability of the population of interest is not assessed. While estimating the correlation of observations within the sample is helpful in identifying lack of independence, it is not enough; they may be conditionally independent. The illustration deals with inadequate replication over time and space.

The essential criterion is to ask whether the samples are fully representative of the population of interest. The term "representative" has been discussed already in connection with Rule 1.1, page (2) . This is, first of all, a qualitative criterion. As indicated, it is a slippery term with intuitively obvious content but difficult to define. In the illustration above, each of the three sampling situations does have a population for which the samples provide valid estimates. So it is first of all a question of defining very carefully the population of interest. The statistical consultant needs to get consensus about the population to be studied before making judgments about the appropriateness of a sampling scheme.

The unit of observation needs to be carefully defined. It will pay to define and list the ways that randomness enters into the selection of the data. Each source of randomness has associated with it a specific unit of observation.

5.5 THINK BEYOND SIMPLE RANDOM SAMPLING

Introduction

Sampling is an essential component of environmental studies. Frequently there are regulatory requirements for sampling, or regulatory implications. This places a premium on valid samples and sampling procedures. The chemical analysis of environmental samples can be very expensive and time-consuming so that samples—and their statistical analysis—should provide as much information as possible. Also, collection of some environmental samples may have substantial occupational hazards so, again, thorough attention to the sampling scheme is important.

Rule of Thumb

Always consider alternatives to simple random sampling for a potential increase in efficiency, lower costs, and validity.

Illustration

Pearson and Rose (2001) discuss the "effects of sampling design on estimating the magnitude and distribution of contaminated sediments in a large reservoir." Besides simple random sampling, they consider random, stratified, systematic or transect, and staged sampling. Given the heterogeneity of contamination of the reservoir, they recommend a staged sampling approach with the first stage being used to determine spatial heterogeneity. The determination of heterogeneity could then be used to develop a stratified design.

Basis of the Rule

Simple random sampling is the simplest form of probabilistic sampling involving a uniform distribution. But the sampling distribution can be chosen by the investigator and alternatives are possible so that the sampling objectives are reached with greater validity and smaller expense. Alternatives to simple random sampling may have greater efficiency associated with them.

Discussion and Extensions

There are many forms of probabilistic sampling. The field of survey sampling provides a useful set of sampling strategies. A good discussion of environmental sampling strategies can be found in Bowen and Bennett (1988) and the more recent book by Millard and Neerchal (2001). The latter is encyclopedic and has the advantage of providing S-Plus routines for all of the statistical procedures mentioned.

In addition to the survey sampling models, two sampling strategies are associated with environmental sampling: composite sampling and ranked set sampling. Composite sampling has intuitive appeal. The first use of it is attributable to Dorfman (1943), who suggested the strategy of pooling blood samples of army recruits tested for syphilis. If the pooled sample was negative, no further testing was done. If positive the original samples were analyzed. If the condition to be tested for is rare, this clearly provides potential savings. In environmental studies this could be applied to sampling for contaminated hot spots. The work by van Belle et al. (2001)

suggests that pooling samples of five provides substantial savings over a wide range of prevalences from 0.005 to 0.05. The cost relative to no compositing is given by the formula

$$\text{Relative cost} = \frac{n+1}{n} - (1-\pi)^n, \tag{5.10}$$

where n is the number of samples in the pooled sample and π is the prevalence of contaminated samples, say. At a prevalence of 0.10 there is a slight advantage to pools of five versus pools of 30. The paper cited also contains a new compositing scheme called double compositing. Composite samples are also taken in order to increase precision of measurements. See the June 2001 issue of *Environmental and Ecological Statistics* for the most recent research in composite sampling.

In ranked set sampling initial, quick, or cheap measurements related to an outcome are used to select samples for subsequent, more expensive and time-consuming analysis (see Mode et al., 1999). Millard and Neerchal (2001) point out that ranked set samples will tend to be more representative of the population being measured. Both sampling strategies provide potential savings and increases in precision.

The key to optimal use of sampling is that simple random sampling is not necessarily the best; as long as the sampling probabilities are known there may be better strategies.

5.6 CONSIDER THE SIZE OF THE POPULATION AFFECTED BY SMALL EFFECTS

Introduction

It is an epidemiological rule of thumb that odds ratios less than two (or greater than 0.5) are not very interesting. In environmental studies this is not necessarily true.

Rule of Thumb

In assessing the importance of an effect, consider the size of the population to which it applies.

Illustration

Low levels of air pollution have rather small effects. But the number of people at risk is huge so that the overall effect can be quite large. For example, the National Morbidity and Mortality Air Pollution Study (NMMAPS) sponsored by the Health Effects Institute of Boston found a "0.5% increase in total non-accidental mortality associated with a $10 \, \mu g/m^3$ increase in PM_{10} in the 90 largest US cities where daily average PM_{10} ranged from 15 to 53 $\mu g/m^3$" (Samet et al., 2000b). PM_{10}, and $PM_{2.5}$ are particles less than 10 and 2.5 micrometers (μm) in diameter, respectively. These are very small effects, but huge numbers of people are affected. Another analysis by Murray and Nelson (2000) of data for Philadelphia, confirming these results, suggests that there is an interplay between temperature and pollution levels and also suggests that the life-shortening effect is approximately two weeks. These kinds of studies, with these kinds of effects, raise many questions about validity of the models—a topic that will be discussed in the next rule.

Basis of the Rule

The rule is based on the effect on the public's health rather than a strict scientific, mechanistic explanation.

Discussion and Extensions

The statistical methodology for detecting very small effects needs to be sophisticated and also requires huge databases. Such databases are often constructed not for scientific purposes but administrative or regulatory reasons. This does not mean that they cannot be used, but a caution flag needs to be in the statistician's mind when using them. For example, the EPA AIRS database measures PM_{10} every sixth day. This means that daily mortality effects cannot be detected with this kind of data.

This discussion also illustrates that the concept of effect size is inadequate, or needs to be redefined, in a public health and policy context. Additional considerations are the number of people involved, costs, and ease of application of remedial techniques.

Sample sizes in studies of small effects are huge. The usual calculations for sample size do not apply to these kinds of studies. Analysis of these small effects requires specialized statistical, computing, and database expertise.

5.7 STATISTICAL MODELS OF SMALL EFFECTS ARE VERY SENSITIVE TO ASSUMPTIONS

Introduction

The use of statistical procedures to estimate small effects affecting large populations places special burden on the statistician to use a valid model.

Rule of Thumb

Models estimating small effects in large populations are particularly sensitive to assumptions. Extensive sensitivity studies are needed in such cases to validate the model.

Illustration

A paper by Lumley and Sheppard (2000) examined model sensitivity to seasonal confounding and selection bias. They found the bias to be small but of similar order of magnitude as the estimated health effect. If there is a substantial health effect the bias can be ignored but in this kind of situation it needs to be weighed carefully.

Basis of the Rule

Complex models have underlying structures that are not always fully transparent to the investigators. Subtle effects due to selection bias and measurement error may mistakenly produce significance of components of the model. Another source of problematic inference is when the cause is multifactorial, and there is no plausible mechanism to identify one or more of the factors as causal. This has been a persistent problem in the assessment of the effects of automobile air pollution on health. Current dogma has identified particulates as a key culprit, but there still is debate about this in view of a lack of mechanism and conflicting results.

Discussion and Extensions

Sensitivity analysis is a standard approach to investigating the impact of assumptions on the model. Such analyses cannot be carried out in a cavalier fashion, they require as much planning as a carefully conducted experiment; they are the "bench science" of the statistician. Samet et al. (2000a, 2000b) did routine sensitivity analyses of their models, underscoring the robustness of their conclusions.

Another arena where small effects need to be assessed carefully is publication bias. An illustration is the assessment of the effect of passive smoking. The putative effect is lung cancer. There is a very long latency period, a knotty exposure assessment problem, and a small effect. Copas and Shi (2000) reviewed the literature and estimated that publication bias could reduce the excess risk from 24% to 15%. They do not deny that there is a risk; their point is that publication bias may have exaggerated the risk.

McMichael et al. (1998) point out that linking air pollution levels to daily mortality cannot produce estimates of long-term mortality. Such effects must be studied by means of cohort studies rather than essentially cross-sectional studies. Using very sophisticated statistical models in this case to detect such small effects is doomed to failure from the start. A non-significant result is to be expected.

The concluding comments to the previous rule apply to this rule as well. This area is not for the statistical neophyte armed with a powerful computing package but no nuanced statistical expertise. The key to assessing small effects is a thorough understanding of the subject area so that causal mechanisms can be translated into a testable statistical context and evaluated.

5.8   DISTINGUISH BETWEEN VARIABILITY AND UNCERTAINTY

Introduction

Statistics is the analysis of variation. There are many sources and kinds of variation. In environmental studies it is particularly important to understand the kinds of variation and the implications of the difference. Two important categories are variability and uncertainty. *Variability* refers to variation in environmental quantities (which may have special regulatory interest), *uncertainty* refers to the degree of precision with which these quantities are estimated.

Rule of Thumb

In assessing variation distinguish between variability and uncertainty.

Illustration

In assessing the PCB levels at a hazardous waste site be sure to distinguish between the geographical distribution of the PCB levels (variability) and the precision of the estimates of the levels (uncertainty).

Basis of the Rule

Different sources and kinds of variation must be taken into account in analyzing environmental studies. Typically, regulators are more concerned with variability, statisticians with both.

Discussion and Extensions

A rough rule is that variability is good, and uncertainty is bad; or, reduce uncertainty but account for variability. According to Cullen and Frey (1999), there are two kinds of uncertainty: model and input. Model uncertainty deals with the appropriateness of the model used in the risk assessment—for example, are the data normally or lognormally distributed. Input uncertainty deals primarily with measurement error. Model uncertainty is often minimized or not even discussed, yet it is probably the key reason for invalid inferences. Model uncertainty can be assessed by sensitivity analyses or embedding the model in a larger class of models indexed by additional parameters.

In the context of setting standards the distinction between uncertainty and variability has played a key role. See the book by Barnett and and O'Hagan (1997) for a discussion in the context of environmental standards in the United Kingdom, and see Guttorp (2000) and Cox et al. (1999) for discussions of ozone standard settings in the United States.

Mixing the assessments of variability and uncertainty results in the error of pseudoreplication, see Rule 5.4. Doing two laboratory analyses of each environmental sample does not double the inferential sample size—it may double the cost but not the inferential information.

Another way of thinking about variability and uncertainty is that different sources and kinds of randomness are involved. Accounting for them is a prerequisite for a valid and comprehensive statistical analysis. See Rule 6.9 page (143), for additional discussion.

5.9 DESCRIPTION OF THE DATABASE IS AS IMPORTANT AS ITS DATA

Introduction

A great deal of environmental data analysis deals with databases—frequently huge ones. As mentioned before, such data bases are frequently constructed for regulatory and/or monitoring purposes. The primary purpose is not for scientific analysis, using them for research purposes constitutes what has been called *secondary analyses*.

Rule of Thumb

In using a data base, first look at the metadata, then look at the data.

Illustration

The AIRS (Aerometric Information Retrieval System) database is a huge EPA database (http://www.epa.gov/airs/) dealing with air pollution measurements at stations located across the country. Before analyzing data from this database, get information about the location of the stations, the reasons for the locations, the frequency of measurements of pollution levels, instruments used, methods of analyses, and so on. Such stations are operated by the states so there is apt to be a great deal of heterogeneity in the way they are run.

Basis of the Rule

The old computer acronym GIGO (Garbage In, Garbage Out) applies to the use of large databases. The issue is whether the data from the database will answer the research question. In order to determine this, the investigator must have some idea about the nature of the data in the database—that is, the metadata.

Discussion and Extensions

Using large databases, such as AIRS, may involve unstated assumptions—for example, to correlate the daily pollution levels as measured in a city with asthma admissions to hospitals implies that monitor readings are representative for the children involved. To begin to address this, it must be known where the monitor is located, when the monitor operates, how the measurements are made, and where the children live. All of this information is metadata. If for example, the monitor averages pollution readings over a day, and the admissions are thought to be associated with acute exposures to the pollutants, the monitor readings may not be very useful.

Statisticians have begun to develop very sophisticated models to estimate the uncertainty associated with the use of central monitors to represent individuals. This is part of the "errors in measurement" research. For an example, see Samet et al. (2000a).

Metadata is particulary important in collaborative studies involving many centers. This is well known in environmental research, and a great great deal of emphasis has been placed on standardization. The federal government of the United States has developed extensive, multi-agency requirements for describing digital geospatial data (see http://www.fgdc.gov/metadata/constant.html). An international standard is being developed.

It is more difficult to summarize metadata because they tend to be more descriptive and non-numerical. However, this does not detract from their importance.

5.10 ALWAYS ASSESS THE STATISTICAL BASIS FOR AN ENVIRONMENTAL STANDARD

Introduction

Standards are the norm in societies. Many of those standards involve measurement. There are standards for highway speed, blood alcohol while driving (note that breath alcohol is a surrogate), *E. coli* in foods, and many others. Government agencies such as the Environmental Protection Agency (in the United States) and Environment Canada, set standards for pollution levels in air, water, and soil. Since environmental pollution levels vary over time and space, these standards deal with stochastic variables. Standards are used to induce compliance, to enforce corrective action, and to detect unacceptable environmental pollution levels (see Millard, 1987a).

Rule of Thumb

It is not always easy to determine the statistical basis for a standard. Be sure to understand the statistical underpinning of the standard.

Illustration

Standards are changed as new scientific evidence is collected, or new administrations take over. Sometimes they are changed due to legal pressures. For example, in the United States, the Supreme Court in 2000 ruled that the Environmental Protection Agency had to provide additional justification for a proposed tightening of standards for ozone; the court did not disagree that the EPA could set the standard, but judged that the scientific rationale had been inadequately documented. The standard was set on the basis of an extensive collection of peer-reviewed articles and subsequent statistical analysis.

Basis of the Rule

Standards are the history of the interplay of science, politics, and values. The science in many cases is based on very expensive epidemiological studies with crucial statistical analyses. The statistical methodology will be scrutinized, and challenged, by groups directly affected by the change in standards. Statistics and statisticians operate in a fish bowl in these situations.

Discussion and Extensions

The EPA's process for setting ozone and particulate standards provides a good example of the fishbowl activity of statistical and epidemiological analysis. It involved science, politics, the courts, and the media. This kind of scrutiny and challenge of scientific research is very different from the model of the independent scientist working in isolation. For an interesting summary see an editorial by Cullen et al. (2000).

Standards are frequently defined to protect sensitive subpopulations. For example, the ozone standard is designed to protect children with asthma. Identifying and characterizing a sensitive subpopulation has many statistical challenges.

Air pollution standards in the United States are designed to protect human health (primary standards) and human welfare (secondary standards). Secondary standards involve aspects such as visibility, odor, and others that do not necessarily involve human health.

Standard setting and standards involve a great deal of statistical background. Statisticians should—but are not always allowed to—play a key role in this work.

5.11 HOW A STANDARD IS MEASURED AFFECTS POLICY, ENFORCEMENT, AND RESPONSE

Introduction

There may be broad agreement that a pollution problem exists, be it in air, water, or soil. There is more diversity of opinion about quantifying—that is, measuring—the pollutants. This diversity occurs in part because the measurement may link more strongly one of the possibly many sources of pollution rather than another.

Rule of Thumb

How a pollutant is measured plays a key role in identification, regulation, enforcement, and remediation.

Illustration

The level of particulates in the air in the United States has been regulated by the Environmental Protection Agency (and its predecessors) for more than 40 years.

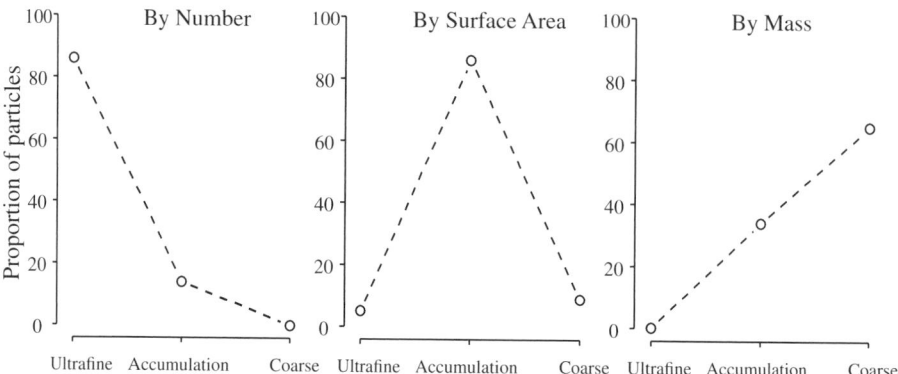

Fig. 5.1   Proportion of types of aerosols expressed by number, surface area and mass. (Based on Table 2 of Whitby (1978); for the continental United States.)

Initial regulations specified total suspended particulates (TSP). Subsequently, PM_{10} and $PM_{2.5}$ were regulated. In all cases it is the mass that is being regulated, not the number, or total surface area of the particulates. One theory about the mechanism of effect of particulates is that there are harmful chemicals on their surfaces so that it should be regulated rather than mass. Of course, there is covariation (not correlation!) between the two so that regulating one will have an effect on the other, but this may not be optimal.

Basis of the Rule

The scale of measurement of a pollutant is important because those regulated will modify emissions to meet this scale. Ordinarily, regulation involves quantification, and quantification necessarily implies reduction of complex systems and sources. A system may be quantified as high by one measure but not by another. A source of pollution may be rated high on one scale and low on another.

Discussion and Extensions

The control of particle pollution in the United States provides a good example of the importance of measurement. To date, as indicated, regulations have involved mass of the particles, not number or surface area. The regulations have consistently reduced the mass, but the number of ultrafine particles ($< 1\,\mu$m in diameter) may have increased. These are the particles that have been implicated in the latest studies of the effect of particulate pollution on health. Figure 5.1 is based on data from Whitby (1978) who analyzed the mix and composition of particles in the continental United

States. This figure shows that in terms of mass, the ultrafine particles contributes the least to an aerosol mixture (panel 3). However, in terms of number, ultrafine particles constitute the largest number. An automobile manufacturer will attempt to reduce mass of particles because the standards are set that way, even by increasing the number of ultrafine particles. (One way this can be done is by moving from gasoline engines to diesel engines which tend to produce more, smaller particles.)

This rule illustrates that the scale of measurement of a standard has practical implications, which may not always be appreciated or anticipated by the standard setters. The process of public hearings on proposed standards can be very useful in elucidating many of these implications. The numerical characteristics of a standard need to be specified carefully keeping in view the purpose of the regulation.

5.12 PARAMETRIC ANALYSES MAKE MAXIMUM USE OF THE DATA

Introduction

Environmental data tend to be heterogeneous, and often the heterogeneity is not well understood or described. The question then comes up, What kind of analysis should be carried out? The two most common candidates are nonparametric and parametric analyses.

Rule of Thumb

Ordinarily, a parametric analysis is the most informative. A nonparametric analysis is useful for an initial, quick look at data (see the next Rule for an example).

Illustration

In a correlation situation the Pearson (product-moment) correlation should be the first choice; the Spearman rank correlation a second. One immediate advantage is that the Pearson correlation can be embedded in a regression analysis involving additional covariates and, perhaps, bringing to the foreground that the covariation investigated is essentially asymmetrical. It is difficult to embed the Spearman statistic in such an asymmetrical inference procedure.

Basis of the Rule

A parametric analysis is usually more informative and permits more complicated modeling. Another drawback of nonparametric analyses is that they usually involve medians rather than means, the latter have more immediate public health applications since they are more explicitly related to the total effect. Finally, interval estimation is more difficult to do in nonparametric analysis; for example, it is not always possible to get interval estimates with specific confidence levels.

Discussion and Extensions

Nonparametric procedures are particularly useful in testing hypotheses. They are less suited to estimation of associations. In the illustration above, the Pearson correlation naturally flows into regression modeling. While a similar approach exists for the rank correlation, the modeling is more complex and less intuitive.

Parametric methods tend to be more informative. Since environmental samples are often very expensive, they have to be used for all they're worth. This means using model-based approaches.

Nonparametric approaches are often associated with dichotomizing data so that discrete distributions such as the binomial can be applied. As discussed in the previous chapter, there is the potential for substantial loss of information, see Rule 4.11, page (99). When associations are involved, the cut point will affect the estimated quantity.

If parametric methods are used, then careful attention should be paid to potential outliers and the form of the distribution.

A parametric analysis involves stronger assumptions than a nonparametric analysis. It is a good statistical consulting principle to begin with the stronger assumption and then, if necessary, weaken the assumptions, see Cox (1999). A drawback of parametric analysis is that the immediacy of the results can be obscured by the intricacy of the statistical model. For example, the results of a factor analysis are difficult to explain in simple terms. The next rule provides an example of immediacy of understanding engendered by a nonparametric analysis.

5.13 DISTINGUISH BETWEEN CONFIDENCE, PREDICTION, AND TOLERANCE INTERVALS

Introduction

Interval estimation can be associated with three types of intervals. The most common is the $100(1-\alpha)\%$ confidence interval. However, there are at least two other intervals that are common and useful in environmental studies.

Rule of Thumb

Distinguish among confidence, prediction, and tolerance intervals. Confidence intervals are statements about population means or other parameters. Prediction intervals address future (single or multiple) observations. Tolerance intervals describe the location of a specific proportion of a population, with specified confidence.

Illustration

A regulatory agency is concerned with the blood lead levels in children. A population-based sample of children is selected and their blood lead levels are measured. The average lead level and confidence interval provide an inference about the population

Table 5.2 Confidence Levels Associated with Nonparametric Confidence, Prediction, and Tolerance Intervals for Samples of Size n Based on Their Extreme Values

		Kind of Interval		
Type	n **observations**	**Confidence**	**Prediction**	**Tolerance**
Lower	(min y, ∞)	$1 - 0.5^n$	$\dfrac{n}{n+1}$	$1 - p^n$
Upper	($-\infty$, max y)	$1 - 0.5^n$	$\dfrac{n}{n+1}$	$1 - p^n$
Two-sided	(min y, max y)	$1 - 0.5^{(n-1)}$	$\dfrac{n-1}{n+1}$	$1 - p^n - n(1-p)p^{(n-1)}$

mean. A prediction interval specifies the range of blood lead levels that a randomly selected child in the population might have. A tolerance interval provides limits within which, say 80% of population blood levels are located, with a specified probability.

Basis of the Rule

The purpose of the study determines the appropriate interval.

Discussion and Extensions

Each of the three kinds of intervals discussed above has two versions—one-sided versus two-sided—and can be estimated parametrically or nonparametrically.

Vardeman (1992) gives a nice discussion of these intervals and illustrates them as follows. The EPA is interested in the fleet average fuel economy of passenger automobiles (point estimate and confidence interval). A car renter is interested in the specific fuel economy for the specific car because it will allow a decision whether a full tank is enough to make a 350 mile trip (prediction interval). A design engineer has to design a tank to make sure that, say, 99% of this type of automobile will have a 400 mile cruising range (tolerance interval). Similarly, environmental protection agencies are often interested in tolerance intervals in order to protect the majority of the population.

A quick confidence interval for the median can be constructed from any sample. Suppose there are n observations in a sample, then the smallest and largest observations form a $\left(1 - 0.5^{n-1}\right)$ confidence interval for the median. These values should be routinely examined in a sample and the associated confidence level ascertained. Table 5.2 (following Vardeman, 1992) summarizes the information in the smallest and largest values in a sample of size n with respect to confidence, prediction, and tolerance intervals. For example, in a sample of size 10 the smallest and largest values form a $1 - 0.5^9 = 99.8\%$ confidence interval for the population median and form a $9/11 = 82\%$ prediction interval for the next observation. If $p = 0.9$ is the fraction of the population to be covered by the interval, there is $1 - 0.9^{10} - 10(1-0.9)0.9^9 = 26\%$ confidence that this interval provides the specified coverage. However, there is 99% confidence that this interval covers 50% of the population. The equations in Table 5.2 can be reversed by specifying the confidence level and solving for the sample size, similar to power calculations.

Corresponding to these nonparametric intervals are parametric ones based on specific models. If the models are correct, the intervals will usually be shorter. However, the nonparametric ones are useful for starters. See Millard and Neerchal (2001), Chapter 6, for details about such parametric intervals. This illustrates that while the nonparametric analysis has intuitive appeal, more exhaustive uses of the data will almost certainly involve their parametric analogues. An automobile manufacturer will insist that the statisticians begin with considering parametric intervals.

Similar issues are raised in regression modelling where confidence intervals can be constructed for the expected value of, say, the intercept or of a future observation of the outcome variable for a given value of the predictor variable. See Kleimbaum et al. (1998) for a detailed discussion.

The two primary aims of statistical inference, estimation and precision of the estimate, are illustrated by this rule. The rule also indicates that the specification of the inferential question must be carefully considered in order to use the correct statistical procedure.

5.14   STATISTICS PLAYS A KEY ROLE IN RISK ASSESSMENT, LESS IN RISK MANAGEMENT

Introduction

Risk assessment has become an essential component of the training of environmental scientists, and statistical methods are an important part of the process. The term *risk* is somewhat ambiguous. It can refer to an adverse event such as dying in a climbing accident, or it can refer to the probability of the event. In most cases it will be clear which is meant. Another term is *hazard*, which has the same kind of semantic ambiguity. Since uncertainty and probabilities are involved, the subject matter is inherently statistical. A fairly standard approach divides risk assessment into five areas (National Research Council, 1983):

1. Hazard identification.

2. Dose–response evaluation.

3. Exposure assessment.

4. Risk characterization.

5. Risk management.

Hazard identification investigates pollutants that have deleterious effects. Dose–response examines the effect of various levels of the hazard. Frequently this is based on animal-based toxicological studies. This presents two statistical challenges. First, there is the problem of extrapolation from animals to humans. Second, since such animals are frequently exposed to very high doses of the putative hazard, there is the problem of high-dose to low-dose extrapolation. Exposure assessment determines who, in the population, is exposed (see the next Rule). Risk characterization integrates the first three steps by defining the risk to the exposed population. These are the key elements for regulators to consider in managing a risk.

Rule of Thumb

The first four activity areas are heavily dependent on statistical input. The last stage involves regulation and policy, based on the information obtained in the first four steps.

Illustration

The identification of radon as the second most important cause of lung cancer (next to cigarette smoking) provides a good example of the risk assessment process. Initial identification was based on fairly high levels of radon exposure in miners. Animal studies, at rather high doses, established a dose-response relationship which then presented the challenge of interpreting these results for human exposures at much lower levels. There was the additional problem that exposures in the animal studies were relatively acute, whereas human exposures are chronic. Exposure assessment then investigated location and levels of exposures in homes. Using all the information above the risk could be characterized. The risk was managed by regulating permissible levels in homes, by developing procedures for mitigating the levels, and by identifying regions in the country with unusually high levels.

Basis of the Rule

The rule is based on the fact that the science of risk assessment is involved in the first four steps, and the fifth is more of an integration of the science with values and politics.

Discussion and Extensions

Since life is inherently risky, the question comes up: What is an acceptable risk? What may be acceptable to one person may not be so to another. Examples are auto racing, smoking, overeating, and so on. There is a plethora of publications and fun books that list and compare risks. Crouch and Wilson (1982) is still readable and quoted widely. A recent contribution is edited by Bate (1997). See especially the paper by Sandman (1997), who has turned the perception of risk into a science. He lists seven principles that characterize media coverage of risk. One of his main points is that the media emphasizes outrage rather than risk, blame rather than hazard, and fear rather than objectivity.

There is increasing awareness that risk assessment involves a sixth stage: accountability. That is, how do the managers and regulators know that the risk management worked? This is a very difficult area because it is usually not possible to count the number of lives saved, or accidents prevented. The cleanest examples of accountability involve traffic–related regulations. The required installation of air bags, for example, provided a very rapid assessment of effectiveness by means of lives saved. Note that in the case of airbags it was also discovered that their use was not completely beneficial due to unexpected deaths of children riding as passengers in the front seat. Ideally, all risk management procedures should have built-in provisions for accountability. The regulation of radon provides a good illustration of the difficulty of accountability; the "number of lives saved" most likely will never be known.

Since the 1980's risk assessment has become a science with its own journal, *Risk Analysis* and a well-developed methodology. The components of risk analysis have been standardized to a great extent with the five steps listed in the introduction.

5.15   EXPOSURE ASSESSMENT IS THE WEAK LINK IN ASSESSING HEALTH EFFECTS OF POLLUTANTS

Introduction

Statistical methods associated with exposure assessment are relatively recent. A paper by Ott (1995b) has the title "Human Exposure Assessment: The Birth of a New Science." Exposure assessment has been around for some time, but Ott explains that the quantitative (read: statistical) methodology of exposure assessment is fairly new—about 20 years at the time of the article.

Rule of Thumb

In the chain from pollution generation to health effects, the assessment of human exposure is the most fragile and most complicated link. Examine the validity of the exposure assessment before doing risk assessment.

Illustration

It has been estimated that 75–80% of all cancers have an environmental cause (Zahm et al., 1995). The chain from pollution generation to occurrence of a cancer is long and tortuous. Characterizing that piece of the chain that deals with human exposure is the most challenging and the least defined.

Basis of the Rule

Exposures and disease are usually separated widely in both time and space. Retrospective quantification of exposure is very difficult—especially if the causes or mechanisms are poorly understood.

Discussion and Extensions

Exposure assessors usually distinguish between acute and chronic exposure. Acute exposure is somewhat equivalent to cross-sectional exposure while chronic exposure has a longitudinal thrust. Acute exposures may be much higher than chronic exposures. Environmental and occupational standards therefore specify limits for both.

New research methods such as biomarkers have not been as productive as originally projected—a good illustration of care to be taken in the use of surrogates. Rule (4.9). Distinctions should be made between biomarkers of exposure, disease mechanism, and effect . They serve very different scientific purposes. For example, biomarkers of exposure and mechanism are usually poor surrogates for clinical outcome. Identification of candidate genes has had success in identifying populations at increased risk.

A great deal of exposure assessment has relied on surrogate exposures such as diet questionnaires. Exposure assessors are particularly concerned about measurement error. A great deal of statistical modeling has focused on assessing the effects of such errors on both validity and precision. Ott (1995a), and Cullen and Frey (1999), provide good starting references.

Retrospective and prospective exposure assessment provide their own challenges, costs, and opportunities. Starting with exposure is fundamental in assessing risk.

5.16 ASSESS THE ERRORS IN CALIBRATION DUE TO INVERSE REGRESSION

Introduction

Standardization of measurements is crucial in environmental studies, especially for regulatory reasons. For example, each day a set of blanks and standards will be run on a machine to standardize the results. Or very precisely known concentrations of pollutants will be run on an optical device to produce a "standard curve." Readings of unknown pollutants are then run against the standard curve, and this curve is used to measure the concentration of a new sample. The optical reading is an *intermediary response* of no intrinsic interest.

Rule of Thumb

Calibration involves inverse regression, and the error associated with this regression must be assessed.

Illustration

A common regression model is often used for calibration,

$$Y = \alpha + \beta X + \epsilon \tag{5.11}$$

where X is the predictor variable and $\text{var}(\epsilon) = \sigma_{y.x}^2$. For a given set of calibration data

$$\hat{Y} = \hat{\alpha} + \hat{\beta} X. \tag{5.12}$$

However, in calibration X is the concentration of interest and Y is the intermediary response. A specific values of $Y = Y_{\text{obs}}$ determines the corresponding value of

$$X_{\text{est}} = \frac{Y_{\text{obs}} - \hat{\alpha}}{\hat{\beta}}. \tag{5.13}$$

This creates a situation with estimation errors in numerator and denominator. If the errors are small relative to environmental variability, they can be ignored. This is often the case. However, the errors should be assessed, and ignorability should also be assessed. Even if the parameters are known exactly, the ϵ term is still in the model.

Basis of the Rule

The variance of X_{est} using the delta method is made up of four terms.

$$\hat{\text{Var}}(X_{\text{est}}) \approx \frac{\sigma_{y.x}^2}{\hat{\beta}^2} + \left(\frac{1}{\hat{\beta}^2}\right)\hat{\text{Var}}(\hat{\alpha}) + \left(\frac{Y_{\text{obs}} - \hat{\alpha}}{\hat{\beta}^2}\right)^2 \hat{\text{Var}}(\hat{\beta}) + \frac{2(Y_{\text{obs}} - \hat{\alpha})}{\hat{\beta}^3}\hat{\text{Cov}}(\hat{\alpha}, \hat{\beta}). \tag{5.14}$$

This complicated equation identifies the sources of variation in calibration. The first term persists even if population values are substituted for the parameter estimates. If β is relatively small, then the variance of X_{est} will tend to be large (the first two terms). Also, if Y_{obs} is far removed from $\hat{\alpha}$ there is an extrapolation problem. These issues are intuitive; the equation provides a quantitative assessment.

Discussion and Extensions

The estimation of errors of ratios of random variables is an old problem. One of the earliest sources was in the context of bioassay. The classical solution is Fieller's theorem; see Finney (1978). For a specific set of data a bootstrap procedure could be used also. See Draper and Smith (1998) and Millard and Neerchal (2001) for a recent treatment of the calibration problem.

The design of calibration assays is an interesting and important arena for statisticians. Some of the design questions related to linear regression discussed in Chapter 3 apply here. The case discussed here, simple linear regression, is the most straightforward. Generalizations involve nonlinear calibration and multivariate calibration.

The delta method does not always work well with error terms with a skewed distribution. Syrjala (2000) gives an example from fisheries. A thorough discussion of the delta method applied to epidemiology can be found in Lachin (2000). The bootstrap method, as described in Rule (1.15) page 25, provides an alternative approach to estimating calibration errors.

Statistical procedures in environmental studies are frequently scrutinized and challenged; more so than in many other areas of application. The statistician working in this area needs to be very competent, to be able to communicate sometimes involved statistical ideas with simplicity and clarity, and to be able to deal with hostile questioning and skepticism about the data, the analysis and the interpretation. All this is to be done with grace, humor, and humility.

6

Design, Conduct, and Analysis

This chapter presents rules of thumb for the design and analysis of studies. The word "studies" is used as a shorthand term for clinical trials and experiments. The rules enunciated in this chapter will apply to both of these groups.

6.1 RANDOMIZATION PUTS SYSTEMATIC EFFECTS INTO THE ERROR TERM

Introduction

The concept of randomization as a basis for the design of studies is attributable to R.A. Fisher (1935). The idea took time getting established. As late as 1940, W.S. Gossett was convinced that if researchers really knew what they were doing, it would be counterproductive to randomize. The importance of randomness as a basis for inference is stressed in this book, beginning with Rule 1.1 on page (2).

Rule

Randomization puts systematic sources of variability into the error term.

Illustration

Suppose the Environmental Protection Agency wants to measure fuel economy in passenger cars. It selects 100 cars at random (just exactly how to do this is not easy) and gets fuel economy estimates from the next tankful. The average of the sample is a valid estimate of the passenger fleet fuel economy and the sample standard deviation provides the basis for an interval estimate. This also is a valid estimate. The variability in fuel economy is a function of many factors such as driving condition, city or highway, type of vehicle, driver characteristics, and other factors. The random selection ensures that the effects of all these factors become part of the random variability.

Suppose now, that a new fuel is to be tested. The data from the 100 car sample is used to estimate variability and equation (2.3) on page (31) is used to calculate the number of cars per treatment needed to determine whether the new fuel improves fuel economy by a specified amount. Suppose that 200 cars per sample are needed. Following the same random selection procedure used above the agency selects 400 cars and randomly assigns half to receive the standard fuel and half to receive the new formulation. This example illustrates that the variability is real, has implications for sample size calculations, and subsequent study effort.

Basis of the Rule

The basis of the rule is the threefold purpose of randomization. First, by randomization, systematic effects are turned into error. Second, there is an expected balance in the assignment of known and unknown factors that might influence the outcome. Third, randomization provides the basis for statistical procedures such as tests of significance, analyses of variance, and confidence intervals.

Discussion and Extensions

This formulation of the rule for randomization is due to Dr. D.B. DeLury of the University of Toronto.

If, in the illustration, the assignment of cars to fuels had not been random, all kinds of objections could have been raised by interested third parties. The nature of the objections would be determined by whether a significant difference had been found or not.

Lack of randomization leads to arm-waving (Rule 1.1). The most common nonrandomized study is the observational study. The lack of randomization in such studies invariably leads to debate about the interpretation of the results. Epidemiologists come closest to making observational studies as clean as possible through, for example, case-control studies. But even in this situation there is debate about the validity and interpretation of results. As was pointed out in the last chapter, a vigorously debated issue is the relationship between air pollution and mortality (Samet et al., 2000a,b). Epidemiological evidence for such effects is based on time series analysis of daily deaths and daily pollution level at central monitoring stations. One reason for the debate is that such data are observational so they lack the compelling feature of a randomized study.

Randomization is essential for valid inference. Observational studies may provide correct answers but the degree of certainty is much less than for randomized studies; see the discussion in connection with Rule 1.1 on page (2).

6.2 BLOCKING IS THE KEY TO REDUCING VARIABILITY

Introduction

While randomization is the key ingredient in the validity of statistical inference, it does have the disadvantage of generating a potentially large error term. One way around this drawback is to establish homogeneous subsets of experimental units (*blocks*) and assigning treatments randomly to the units within these subsets.

Rule of Thumb

By randomly assigning treatment to experimental units within blocks (i.e., blocking), systematic effects between blocks can be eliminated with a resultant decrease of within-treatment variability and increased precision.

Illustration

In the previous illustration some of the variability in fuel economy was attributable to driving conditions, city or highway. If comparisons are only made within cities, the standard deviation of fuel economy among the city-driven cars will be smaller than the overall standard deviation. Hence, for sample size calculations a smaller sample size will be needed to establish the significance of the putative fuel improvement. The factor, driving condition, is said to be a *blocking factor*.

Basis of the Rule

The total variability in a population can be divided into two components: among blocks and within blocks. By assigning treatments within blocks, between block variability is eliminated. If the background variability in the blocks is less, that is the blocks form homogeneous units, then by blocking more precise estimates are obtained.

Discussion and Extensions

The term *blocking* has an agricultural origin where treatments were compared within blocks of land. Each block was partitioned into "plots" upon which the treatments were carried out. In clinical trials, animal studies, or other investigations the blocks can be people, animals, treatment sessions, cities, school districts, or any group that is more homogeneous than the general population. This applies to observational studies as well.

The "paired study" is a good example of blocking. For example, in comparing the effectiveness of two treatments for epilepsy, one design randomly allocates treatments to groups of subjects with epilepsy and compares, say, seizure frequency after 12 weeks of treatment. The error term then would be the variability in response among the subjects within the treatments—between-subject variability. An alternative design tests both treatments on the same subject. The sequence of treatments is allocated randomly (the source of randomization), and comparisons are within subjects. A paired study is perhaps the most extreme of blocking; all between-subject variability is removed in this design. Blocking is not always efficient as the example associated with Rule 3.5, page (61) indicates.

There is a simple test for the effectiveness of blocking: the variability within blocks must be less than the variability between blocks. This can be assessed by means of an analysis of variance or, in the case of paired data, by correlating the pairs of observations as discussed in Rule 3.5.

Any factor that co-varies with the outcome variable may be a candidate blocking factor. It should not be in the causal chain between treatment and outcome.

The question may raised, why not make blocks as fine as possible? The answer is that there is a trade-off with generalizability of the results.

The concept of blocking is intuitively appealing. Blocking occurs in many situations. In multi-center clinical trials each clinic may be considered a block and treatments are assigned randomly to subjects within the clinic so that there is balance at the clinic level. This eliminates clinic variability from treatment comparisons.

6.3 FACTORIAL DESIGNS SHOULD BE USED TO ASSESS JOINT EFFECTS OF VARIABLES

Introduction

There is an old scientific "rule of thumb" recommending investigation of one factor at a time. Consider the following quote from *Byte* magazine (Halfhill, 1995) which summarizes that view as well as any:

> ... In any scientific investigation, the goal is to control all variables except the one you're testing. If you compare two microprocessors that implement the same general architecture on the same process technology and both chips are running the same benchmark program at the same clock speed, any difference in performance must be due to the relative efficiencies of their microarchitectures.

The problem with this prescription is that in most studies there are many factors that are of interest, not just one factor. The challenge is then to develop a design that allows assessment of all of the factors with minimal additional cost.

Table 6.1 Example of Factorial Design. Entries in the Table Are the Numbers of Subjects

	Ozone	Air	
Active	$n/2$	$n/2$	n
Rest	$n/2$	$n/2$	n
	n	n	$2n$

Rule of Thumb

The effects of two or more factors can be assessed simultaneously by means of factorial designs in which treatment combinations are applied to the observational units. A benefit of such a design is that there may be decreased cost.

Illustration

Suppose that a comparative study has been designed to study the response of asthmatic subjects to two factors: pollutant exposures (ozone and air) and two exercise levels (active and rest). Assume that it has been decided to test n subjects at each level of the two factors. Following the advice given in the introduction, the design involves two separate studies: one study to look at the effect of pollutant and the other study to look at the effect of exercise. This would take $4n$ subjects. An alternative approach is the factorial design which assigns $n/2$ subjects to each of the four treatment combinations as displayed in Table 6.1. This design requires only $2n$ subjects as compared with the $4n$ subjects in the two, separate, single factor studies. Comparison of the means in the bottom row of the table reflects the effect of pollutant; a similar comparison of the column margin reflects the effect of exercise. Hence, this experiment contains the same information as the two independent experiments. In addition a comparison of the means within the table, the cell means, allows examination of the question whether the effect of pollutant is the same during exercise and during rest.

Basis of the Rule

The variance of the differences (comparing active and rest, or comparing ozone and air) is based on the same number of subjects as the single factor study. The precision is virtually the same as that for the two independent studies (see below); but with half the number of subjects.

Discussion and Extensions

It is appropriate to contrast the above quote with another one by Fisher (1935):

> No aphorism is more frequently repeated in connection with field trials, than that we must ask Nature few questions or, ideally, one question at a time. The writer is convinced that this view is wholly mistaken. Nature, he suggests, will best respond to a logical and carefully thought out questionnaire; indeed if we ask her a single question, she will often refuse to answer until some other topic has been discussed.

What did Fisher mean? Essentially that researchers are often interested in relationships between factors, or interactions. For example, it could be that the effect of ozone exposure can only be seen when exercising—a two-factor interaction. Comparison of the means in the margins, as indicated, reflect the effects of the factors separately. These are called the *main effects*. Thus, the factorial design provides not only the information of the two separate studies but also the potentially more important information about the relationship between the two factors.

It is useful to consider the error terms for the single factor study with n subjects per treatment, and the factorial design. In the single factor study the error term has $2n - 2$ degrees of freedom since each treatment has n subjects with $n - 1$ degrees of freedom per treatment level. In the factorial experiment each cell contributes $n/2 - 1$ degrees of freedom to the error term for a total of $2n - 4$ degrees of freedom. So two degrees have been lost. Where did they go? One of them went to the comparison of the other factor, and the other is used in estimating the interaction. Table 6.3 illustrates the analysis for $n = 20$ subjects per treatment (when that there are 10 subjects per cell.

Given this rule, why are not all studies factorial? If there are many factors at many levels, it will not be possible to carry out the full factorial analysis. For example, if there were four factors, each of three levels, then $3^4 = 81$ observation units are needed for only one run of study. (But see the following rule of thumb for a possible solution.) Note also that the precision of the estimate of interaction is only half of that of the main effects. This means that the interaction effect must be large in order to be able to detect it in a factorial design.

The three strategies of randomization, blocking, and factorial design are the basic elements of experimental designs. Their creative and judicious use will lead to efficient, robust, and elegant investigations.

6.4 HIGH–ORDER INTERACTIONS OCCUR RARELY

Introduction

An interaction of factors requires specific mechanisms. For example, suppose that the asthmatic subjects, of the illustration in the previous rule, had shown a pollutant effect during exercise but not during rest. This would require an explanation by a respiratory physiologist, who might argue that pollutants only have an effect when the subject is stressed; as in exercise. This would be a superficial explanation which should lead to the question, "Why would that be the case?" A deeper explanation might then mention the immune system and how it responds to pollutants.

The more complicated an interaction, the more complicated the mechanism of the organism's response. It is an empirical observation that such situations are relatively rare.

Rule of Thumb

High–order interactions occur rarely, therefore it is not necessary to design experiments that incorporate tests for high–order interactions.

Illustration

Consider a study of decline in memory among blacks and whites (race), males and females (sex), ages 70 to 90 (age), and varying levels of schooling (education). There are four explanatory variables (race, sex, age, and education). A full model would involve four-factor interactions. It is difficult to imagine a mechanism that would produce such an interaction. Note also that decline in memory itself is a change variable. If such an interaction were found in an observational study, the investigators would almost certainly explore selection biases and other reasons for this observation. If they were satisfied that the interaction was "real," they would begin to explore mechanisms.

Basis of the Rule

The basis for the rule is that it is difficult to picture mechanisms associated with high-order interactions, particularly in studies involving organisms. There is parsimony of mechanisms. Another consideration is that to a first order approximation, effects of factors are additive. Finally, interactions are always expressed in a scale of measurement and there is a statistical theorem that says that a transformation of the scale can be made so that effects become additive (Scheffé,1959). In other words, some interactions are simply associated with the scale of measurement.

Discussion and Extensions

The existence of an interaction implies that the effects are not additive. Relative risks and odds ratios essentially estimate two-factor interactions; the risk of disease among the exposed compared with this risk among the non-exposed. On the log scale, if there is no interaction the log of the relative risk and the log of the odds ratio are exactly zero, so that relative risk and odds ratio are 1.

Fractional factorial designs allow the application of the two rules of thumb listed above (Rule 6.3 and Rule 6.4). They take advantage of factorial structure while using the higher–order interactions for estimating the error.A classic book discussing this and other issues has been updated recently; see Cox and Reid (2000). It is part of scientific insight to know which interactions can be ignored and which should be considered in the design of a study.

The assumption that high–order interactions do not occur, commonly underlies a great deal of research; it forms a frequently unstated context. It is perhaps common because it is often correct. However, it will certainly pay to itemize the more common high–order interactions assumed to be non-existent.

6.5 BALANCED DESIGNS ALLOW EASY ASSESSMENT OF JOINT EFFECTS

Introduction

Since designed studies are under the control of the investigator one choice is the allocation of resources to the treatment conditions. It turns out that "nice" things happen when the allocations are arranged in a specific way. A sufficient condition for a *balanced design* is that all treatment combinations are based on the same number of observations.

Rule of Thumb

Aim for balance in the design of a study.

Illustration

In a factorial design aim for equal numbers of study units to each treatment combination. This makes the various types of analyses equivalent in statistical packages. For example, with balanced data the Type I and Type III analyses in SAS are identical. Table 6.1 illustrates such a balanced design. However, see the discussion for a more generally balanced design illustrated by Table 6.2.

Basis of the Rule

This rule is based on the resulting analyses that are more straightforward and, typically, allow additive partitioning of the total sums of squares. Studies that are balanced are often called *orthogonal* because additive partitioning of the sums of squares is equivalent to an orthogonal partition of treatment response in the outcome space.

Discussion and Extensions

The concept of balance is fairly straightforward: equal allocation of samples to treatments. However, it need not be quite so restrictive. For example, in a 2×2 factorial design the total sum of squares can still be partitioned additively when the cell frequencies are proportional to the marginal frequencies. To determine whether this is the case, simply calculate the chi-square statistic for contingency tables on the cell frequencies. If the chi-square statistic is exactly zero, the cell frequencies are in

Table 6.2 A Balanced Factorial Design. Entries in the Table Are the Numbers of Subjects

	Ozone	**Air**	
Active	12	8	20
Rest	24	16	40
	36	24	60

proportion to the marginal frequencies. Table 6.2 illustrates such a design. In this design the cell frequencies are determined by the marginal frequencies: $n_{ij} = n_{i.} \times n_{.j}/n_{..}$ for $i = 1, 2$ and $j = 1, 2$. Table 6.5, page (140) also contains data that are balanced.

As indicated, balanced studies provide for additive partitionings of the total sums of squares. Thus there is no need to worry about the sequence of hypotheses that are analyzed. Historically there has been a great deal of emphasis on balanced experiments because the computations were much simpler. This, of course, is less of a problem now. However, there is an additional reason for striving for balanced studies: ease of interpretation. Designs such as balanced incomplete block designs, and partially balanced incomplete block designs, were devised to be relatively easy to analyze and easy to interpret. Another nice feature of balanced allocation is that ordinarily the standard errors of the estimates will be minimized given that the total sampling effort is fixed.

For a one-way classification the numbers per treatment can be arbitrary in that it still allows an additive partitioning into treatment and error sums of squares. However, unless the frequencies are balanced, it will not be possible to subdivide the treatment sum of squares additively (of course, if there are only two treatments the further partitioning is not an issue and the sample sizes can be unequal).

Balance is important for additive partitioning of sources of variability. It is of less importance for precision, as Rule 2.9 page (45) indicated. If balance is required, the imbalance is not too great, and there are a reasonable number of observations per factor combination, it would be possible to randomly discard a few observations in order to obtain balance. If this is done several times (akin to bootstrapping) the average of the results can be taken as a good summary of the analyses.

6.6 ANALYSIS FOLLOWS DESIGN

Introduction

It is not always clear what analysis is to be carried out on the data. One of the key considerations is to determine how the randomization was done since it provides the basis for the tests of significance and other inferential procedures.

Table 6.3 Design A. Analysis of Variance Table for Forty Subjects Assigned Randomly to One of Four Treatment Combinations

Source of Variation	Degrees of Freedom
Pollutant	1
Exercise	1
Pollutant×Exercise	1
Error	36
Total	39

Rule of Thumb

Analysis should follow design.

Illustration

Consider the two following designs assessing the effects of pollution and exercise on asthmatic subjects. In Design A, 40 asthmatic subjects are assigned randomly to one of the 4 treatment combinations. In Design B, each of 10 asthmatic subjects receives all 4 treatments in random order; thus each subject constitutes a randomized block Both designs will generate 40 observations. But the analyses are very different. The analysis of variance table for Design A is displayed in Table 6.3. The design is a completely randomized design with the four treatment combinations randomly assigned to subjects; the randomization is carried out in such a way that equal numbers of subjects are assigned to the four treatment combinations.

Design B is a type of repeated measures design with all treatment combinations assessed within each subject. The analysis is quite different as indicated in Table 6.4.

Table 6.4 Design B. Analysis of Variance Table for Forty Subjects Assigned Randomly to One of Four Treatment Combinations

Source of Variation	Degrees of Freedom
Between Subjects	9
Pollutant	1
Exercise	1
Pollutant×Exercise	1
Error	27
Total	39

Basis of the Rule

The sampling strategy used to generate the data determines the appropriate analysis—tying the two together tightly.

Discussion and Extensions

The difference between the two analyses is most clearly seen in the degrees of freedom for the error terms. In Design A, the error term has 36 degrees of freedom, based on 9 degrees of freedom for each of the cells. It is a between-subject error term. In Design B, 9 degrees of freedom are subtracted to account for between-subject variability leaving 27 degrees of freedom for the error term. This is a within-subject error. Ordinarily, this error term will be smaller than the error term for Design A, so there will be greater power to detect a treatment effect.

Violations of the rule usually affect the error term but not necessarily the estimate. For example, analyzing paired data as two independent data sets still provides a valid estimate of the difference but an invalid estimate of the precision of the estimate.

There are subtle issues that are not always recognized. It is incorrect to analyze a two-factor design involving one factor a treatment factor, and the other a classification factor (for example, high school education) as a factorial design since education is not randomly allocated to the subjects. Such an analysis may be valid, but will require some of the hand-waving discussed in Rule 1.1, page (2), and Rule 6.1, page (129). Technically, the treatments are nested within education.

Just as sample size calculations should be based on the proposed analysis (Rule 2.12, page 50), the statistical analysis should follow design. These rules indicate the tight linkage between analysis and design; both force the investigator to look forward and backward.

6.7   PLAN TO GRAPH THE RESULTS OF AN ANALYSIS

Introduction

As will be discussed in the next chapter, statistical packages usually do not do a very good job of presenting graphical analyses of data.

Rule of Thumb

For every analysis there is an appropriate graphical display.

Illustration

Nitrogen dioxide (NO_2) is an automobile emission pollutant. Sherwin and Layfield (1976) studied the protein leakage in the lungs of mice exposed to NO_2 at 0.5 parts per million (ppm) NO_2 for 10, 12, and 14 days. Half of a total group of 38 animals was

Table 6.5 Serum Fluorescence Readings of Mice Exposed to Nitrogen Dioxide (NO_2) For 10, 12, and 14 Days Compared with Control Animals: Means, Standard Deviations and Sample Sizes

			Exposure		
			10 Days	**12 Days**	**14 Days**
	Control	Mean	134.4	143.0	91.3
		S.D.	24.7	35.5	43.2
		n	7	6	6
Pollutant					
	Exposed	Mean	106.4	173.2	145.5
		S.D.	32.1	51.0	37.1
		n	7	6	6

exposed to the NO_2; the other half served as controls. This is a two-factor factorial design with one factor (Pollutant) at two levels, and the other factor (Days) at three levels. The means, standard deviations and sample sizes are given in Table 6.5. The analysis of variance is presented in Table 6.6.

Since the controls are intended to adjust for day-day variation in response, it is appropriate to subtract the control mean from the treatment mean on each day. This provides the basis for the test of interaction of treatment and days. Figure 6.1 demonstrates that there is an increasing NO_2 effect with days. The vertical bars are twice the associated standard error. In the right-hand panel of Figure 6.1 this bar indicates that the trend is significant and not inconsistent with a linear trend in treatment effect (as compared with control). A partition of the the interaction mean square in Table 6.6 confirms this impression (not shown). For further discussion of the Sherwin and Layfield (1976) study see Fisher and van Belle (1993).

Table 6.6 Analysis of Variance Associated With Serum Fluorescence Data of Table 6.5

Source of Variation	d.f.	M.S.	F	p
Treatment	1	2528.95	1.78	>0.10
Days	2	6124.48	4.32	<0.05
Treatment×Days	2	5873.6	4.14	<0.05
Residual	32	1417.35	—	—

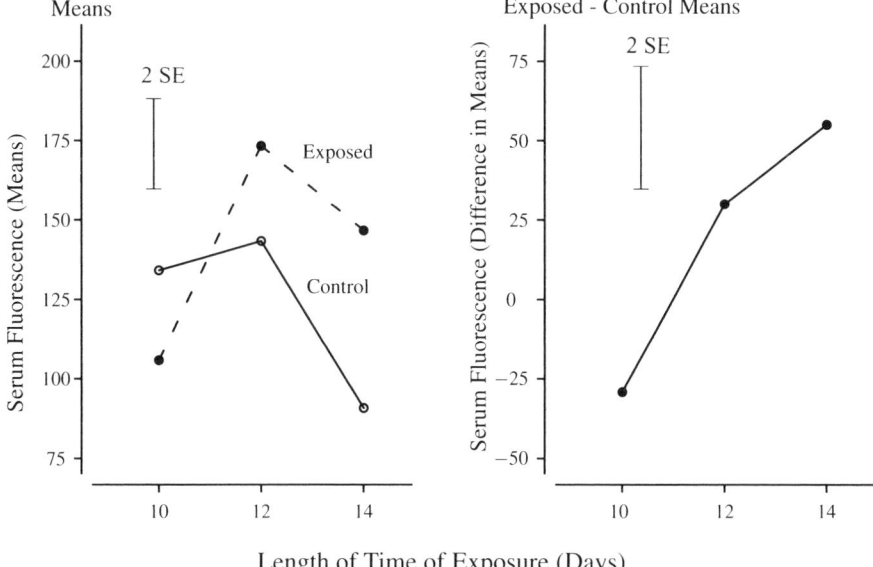

Fig. 6.1 Graphical display associated with analysis of variance in Table 6.6. Twice the standard error of the Mean for $n = 7$ is 28.4, for $n = 6$ is 30.8. Twice the standard error of the difference of the Means for $n = (7, 7)$ is 40.1, for $n = (6, 6)$ is 43.4.

Basis of the Rule

Every statistical analysis involves estimates, relationships among the estimates, precision of these estimates and, possibly, covariates. These features can all be displayed pictorially. Figure 6.1 clearly illustrates where the effects are and why the the terms are significant.

Discussion and Extensions

In the case of treatment comparisons use interaction plots rather than bar graphs. A row in an analysis of variance table with, say $k - 1$ degrees of freedom involves a comparison of k quantities. These quantities can be meaningfully graphed to illustrate the effects and their significance. Factorial designs are particularly easy to graph. Interactions involve some kind of non-parallelism and this can be nicely illustrated in a variety of ways. The left panel of Figure 6.1 displays the interaction in terms of lack of parallelism; the right panel shows it by means of a non-horizontal line (since the ordinate is a difference, and a change in the differences represents interaction). Such graphs also are useful for pointing out outliers or unusual patterns. There is room for creativity here.

Figure 6.1 also illustrates the importance of the hierarchical analysis principle of looking at the highest order term first. The overall mean difference between exposed and control (averaged over the three days of exposure) is 16.3 with standard error 12.2, clearly not significant. (Note that $(16.3/12.2)^2 = 1.78$, the value of the F statistic in Table 6.6). This value and its associated standard error have been graphed in the second panel of Figure 6.1. But the interaction is significant so that is where the discussion should start, not with the main effects.

This rule illustrates the importance of thinking graphically about the interpretation of the results of a study. For more rules of thumb for graphs see the next chapter.

6.8 DISTINGUISH BETWEEN DESIGN STRUCTURE AND TREATMENT STRUCTURE

Introduction

There is a great deal of diversity in studies and it is useful to categorize them in a variety of ways. The *design structure* is the physical layout of the study. For example, compare two weighings of each of ten animals with the weighing of twenty animals. In each case there are twenty data points but the design structure is different—and the analysis will differ. The design structure must be known to correctly interpret the results and their implications. The *treatment structure* of a study describes the possible relationships among factors. For example, a factorial treatment structure will involve the application of all possible factor combinations (or fractions thereof) to experimental units.

Rule of Thumb

Distinguish between the design structure and the treatment structure of a study.

Illustration

The introduction to this rule presented a simple example. A richer example is given in Rule 6.6 on page (137), with a factorial design (treatment structure) with each treatment combination assigned to different subjects or to the same subjects (design structure). In that example the treatment structure was the same but the design structure of Design A differed from Design B. The analyses in Tables 6.3 and 6.4 reflect the differences in design structure in the degrees of freedom for the error term, while there are identical degrees of freedom for the treatment structure.

The example discussed in the previous rule has a factorial treatment structure, the design structure involves randomly allocating animals to the treatment combinations— identical to what would be done if the study involved six unrelated treatments.

Basis of the Rule

The design structure determines the layout of the analysis. The treatment structure determines the finer partitioning of the degrees of freedom.

Discussion and Extensions

The concepts of treatment and design structure have proven to be helpful in thinking about the development of designs for studies. They indicate that there are constraints imposed by the study units and the treatment units. Suppose that in the example above it was not possible to assign all treatment combinations to the same subjects (possibly because of carry-over effects or a practical problem of not being able to get the subjects to come in four times). The treatment structure is still of interest but the design structure restricts the experimental design. A great deal of work has been done to make treatment structures and design structures compatible. For example, in factorial studies it may not be feasible (because of cost or time constraints) to study all possible factorial combinations—a four factor experiment with three levels per factor requires $3^4 = 81$ experimental units per replication. This leads to consideration of fractional factorial designs mentioned in Rule 6.4 on page (134). See Cox and Reid (2000) for examples and further discussion. The treatment structure is the driver in planning experiments, the design structure is the vehicle.

Design and treatment structures in experimental designs often require trade-offs. It is important to understand the differences between these structures, to know about possible compromises, and to be able apply the best statistical strategies to the design and analysis of the proposed research.

6.9 MAKE HIERARCHICAL ANALYSES THE DEFAULT ANALYSIS

Introduction

This section assumes a factorial treatment structure with each factor at two levels such as presence or absence, low dose or high dose. Treatment effects in this kind of study can be classified as main effects (attributable to a single factor) or interaction effects (attributable to two or more factors). The interaction effects can be sorted into two-way, three-way, up to k-way interactions if there are k factors. An effect is of higher order if it involves more treatment combinations. With k factors there are 2^k possible models to consider—very quickly a large number. The following rule gives a rationale for assessing a smaller number of models. In addition, it provides a guide to the ordering of the effects in the analysis.

Rule of Thumb

Ordinarily, plan to do a hierarchical analysis of treatment effects by including all lower order effects associated with the higher order effects.

Illustration

Consider a two-factor design consisting of factors A and B. Each factor can be present or absent. The possible treatment effects are A (that is, the effect of A present compared with A absent), B (present or absent), and A×B (both A and B present or both absent). If each of these can be in or out of the model there will be $2^3 = 8$ possible models. If there are 3 factors, A, B, and C, there are $2^7 = 128$ possible models to analyze. The rule states that only hierarchical models should be considered. For example, a model consisting solely of the interaction A×B is not allowed. If A×B is to be examined, the rule states that A and B should also be in the model. Beside the null model (no treatment effects), in the three-factor example, there are 18 hierarchical models—a substantial reduction from 128. Not all of these 18 models are of equal interest. For example, one of the 18 models involves only the factors B, C, and the interaction B×C. Why analyze only these two factors when A is also in the experiment? If the decision is made that only hierarchical models using all factors will be examined then only 9 models need to be assessed.

Basis of the Rule

There are three bases for the rule. First, the concept of parsimony discussed in Chapter 1 in connection with Ockham's razor (Rule 1.5, page 11) applies here. Second, a hierarchical analysis is more straightforward. Third, the interpretation will be less convoluted.

Discussion and Extensions

One rule sometimes stated is that if the interaction in a hierarchical model is significant then there is no need to examine the associated main effects. The basis for this argument is that the presence of interaction implies differential effects of one factor at the different levels of a second factor. There is merit to this rule but there may be situations where there is both an interactive effect and a main effect, and it may be interesting to compare the magnitude of the two effects. In Figure 6.1 this would correspond to the experimental curve substantially separated from the control curve. While this is a possibility the question still comes up, what would this mean?

Bryk and Raudenbush (1992) provide a thorough review of hierarchical models and the advantage of using this analytic strategy.

Interpreting experimental results is difficult. Using the hierarchical inference principle provides an initial simplification and guidance through the mass of data to be analyzed. If a non-hierarchical analysis is contemplated, there should be a deliberate justification for this approach.

6.10 DISTINGUISH BETWEEN NESTED AND CROSSED DESIGNS—NOT ALWAYS EASY

Introduction

The concept of nesting is best understood by means of an example. Consider students within a class and classes within a school. The two observational units form a hierarchy. It would be possible to apply treatments at different unit levels. For example, students within classes could be randomly assigned to different reading programs. In addition, at the class level there could be a comparison of length of time of instruction. One set of classes has one 30 minute period, another set of classes splits instruction into two fifteen minute periods (it would be difficult to assess the second factor within a class). The treatments are aimed at two different observational units: in the reading programs the student is the observational unit, the classroom is the observational unit for the lengths of instruction. This is an example of a nested design. Crossed designs are ones where all treatment combinations can be assigned to the observational units.

Rule of thumb

Distinguish between nested and crossed designs. The analyses will be quite different.

Illustration

The introduction contained an example of a nested design. Another example is the following. Two varieties of plants are to be compared with respect to growth. Since conditions in a greenhouse differ substantially one plant from each variety is put in a pot. There are 30 pots, hence 30 plants of variety A and 30 plants of variety B. Two kinds of fertilizer are used in the pots so that there are 15 pots with fertilizer F1, and 15 pots with fertilizer F2. This is a nested design with varieties nested within fertilizer. The basic unit for the analysis of variety is the plant, the basic unit for the analysis of the fertilizer is the pot. The question comes up, what is the basic unit for the interaction of variety and fertilizer? The answer is: the plant.

A factorial design is a good example of a crossed design.

Basis of the Rule

The randomization determines the analysis, as indicated in Rule 6.6 page (137). The randomization patterns in crossed and nested designs are quite different. The nested design typically involves more than one randomization, applied to different observational or experimental units.

Discussion and Extensions

The analysis of the nested data in the illustration will involve two error terms, one associated with variability within pots and used for comparing variety and the interaction of variety and fertilizer. The second error term estimates variability among pots and is used to compare the effect of fertilizer. These are almost separate analyses.

In some nested designs there is the potential for heterogeneity of variance. For example, if two methods of clean-up are compared in heavily polluted sites and less polluted sites it will almost certainly be the case that the variability will be larger in the heavily polluted sites. Treating data from this kind of study as if it were factorial is methodologically incorrect and has the potential of misleading inferences. Nesting always raises a question of comparability since there is restricted application of Rule 6.1.

It is not uncommon to see experiments involving gender treated as factorial. It is incorrect to assume that, say, a comparison of the effectiveness of two types of reading programs between boys and girls is a factorial design (one factor, the reading program, the other, gender). Data from such a study are often analyzed as if they were factorial. But they are nested, since treatment is nested within gender. Heterogeneity of variance is of concern in this case also.

Nesting is essentially a characteristic of the design structure of the study. The nesting defines the different types of observational units. Treatments can then be applied to the different types of units. The nested design structure determines the nested treatment structure.

6.11   PLAN FOR MISSING DATA

Introduction

Missing data are a fact of research life. In study design missing data becomes an important issue because balance can be destroyed. This then becomes an issue in the analysis. A great deal of methodological work deals with the treatment of missing data. The concept of missing data is intuitive: data that should be present but for some reason or another are not. And there's the rub: Why are the data missing? The last 20 years have seen very creative statistical approaches to dealing with missing data.

Rule of Thumb

The best rule is: Don't have any missing data. Unfortunately, that is unrealistic. Therefore, plan for missing data and develop strategies to account for them. Do this before starting the study. The strategy should state explicitly how the type of missingness will be examined, how it will be handled, and how the sensitivity of the results to the missing data will be assessed.

Illustration

In a study of life-time dental X-ray exposures, missing data is particularly prominent. Respondents may not recall a dental visit, may not recall the type of X-ray taken, and the further back in time the more tenuous the memory. One strategy for assessing missingness is to request permission to collect dental records from a subset of dentists and calibrate the respondents. This will introduces additional sources of missing data such as records no longer available, incomplete records by dentists, or dentists not remembered; but it will also allow some calibration and categorization of respondent recall. For example, do respondents remember full-mouth X-rays better than bite wings? A sensitivity analysis can be planned by postulating a statistical model for discrepancy between remembered X-rays and actual X-rays and the effect on estimates of usage.

Basis of the Rule

Every statistical analysis is an interpretation of the data, and missingness affects the interpretation. The challenge is that when the reasons for the missingness cannot be determined there is basically no way to make appropriate statistical adjustments. Sensitivity analyses are designed to model and explore a reasonable range of explanations in order to assess the robustness of the results.

Discussion and Extensions

Current statistical terminology defines three kinds of missing data:

- *Missing completely at random* or MCAR. For example, a visit is missed because of a snow storm; or a machine breaks down so that a sample cannot be analyzed. This is the most benign situation; it means that the available data can be analyzed without worrying about bias. Analyzing only units that have complete data is valid (a complete case analysis), although there will be loss of precision. There is little to recommend this approach since better options, making use of all of the data are available.

- *Missing at Random* or MAR. Data are missing but the missingness can be explained on the basis of observed variables. For example, in a multi-center trial one of the centers uses equipment that breaks down more frequently than that of the other centers. The missingness is not completely at random, but conditioning analyses on the centers will take care of the problem. It is not necessary to model the missingness. This example illustrates that the nature of missingness is not unique. From the point of view of the specific center the missingness is MCAR, from the point of view of the multi-center study the missingness is MAR.

- *Non-ignorable* missingness. This occurs when the study outcome is selected in some way. A good example is the healthy worker effect in epidemiological studies. To compare, lung cancer mortality rates of coal miners with say age, sex, and race adjusted population rates ignores the fact that only healthy workers will have been selected initially. The selection process is not known so it is difficult to adjust the mortality rates. Another example is censoring in a clinical trial. Sicker patients may tend to drop out because of treatment side effects, or conversely, healthy patients may drop out because they move to a warmer climate. In all these cases modeling is necessary to make the groups comparable.

The most important actions in dealing with missingness are to document it and account for it. If subjects withdraw from a study, are the sicker ones more likely to so? Is the withdrawal related to the treatment? For example, the experimental drug has more severe side effects.

Some of the strategies for dealing with missing data include complete case analysis (already mentioned), simple imputation such as assigning the mean of the observed values for the missing values, multiple imputation involving repeated sampling from the observed values to impute a series of values for the missing data. The imputation procedures work well with MCAR and MAR situations. But their performance in the non-ignorable case is more problematic. Two ways of assessing non-ignorable missingness are the worst case analysis and some kind of sensitivity analysis. For concreteness consider a randomized clinical trial involving a treatment and a placebo. If the dropout rate is higher in one of the two groups there is almost certainly non-ignorable missingness. In the worst case analysis, the worst response from the treatment group is assigned to all the missing values in the treatment group. Similarly, the best response from the dropouts in the placebo group is assigned to all the dropouts in the placebo group. This is a very conservative strategy that researchers may be loath to take, leading them to consult a statistician who may have more creative ideas Such an idea is a sensitivity analysis, similar to the one discussed in Rule 5.7 on page (112). This involves statistical creativity and a thorough understanding of the subject matter. A range of scenarios will be investigated with the purpose of determining the robustness of the analysis of the the study. If the results hold up, the skeptic should be convinced by the thoroughness of the sensitivity analysis.

The missing data terminology is due to Rubin (1976). A formal treatment of the topic can be found in Little and Rubin (1987) with particular emphasis on the analysis of observational data. Little (1998) provides an excellent review and updated introduction. A discussion in the context of clinical trials can be found in Murray (1998), and an epidemiological overview in Vach and Blettner (1998). A very nice epidemiological application can be found in Rivara et al. (2000). Imputation procedures have been developed to a high degree of sophistication and incorporated into computer packages. For a recent reference see Horton and Lipsitz (2001) and references to the literature.

In many cases missingness will be related to one or more of the study variables. Identification of these variables is the first step on the way to adjusting for the missing variables. If the mechanism for the missing data is not known, it is intrinsically impossible to adjust for the missing data.

6.12 ADDRESS MULTIPLE COMPARISONS BEFORE STARTING THE STUDY

Introduction

The "multiple comparison problem" involves the repeated testing of a series of hypotheses and the resultant increasing probability of a Type I error. The fundamental result is that assuming k independent null hypotheses tested at level of significance α, the probability of at least one null hypothesis being declared significant is:

$$\text{Probability of at least one significant null hypothesis} = 1 - (1 - \alpha)^k. \qquad (6.1)$$

This is a non-trivial price to pay. Even for $k = 3$, using an α level of 0.05 the probability is 0.14—almost triple the value of 0.05. So what can be done about this problem?

Rule of Thumb

Develop a coherent strategy for dealing with multiple comparisons before starting a study. The strategy should include consideration of exploratory versus confirmatory research, primary versus secondary endpoints, and final analysis versus interim analyses.

Illustration

In a clinical evaluation of the effectiveness of a new drug, lasting three years, it is decided to do two *interim analyses*. Because the overall α level of 0.05 is to be maintained the researchers decide to "spend" a little of the α level at each of the analyses. They decide to test the interim results at the 0.001 level and the final analyses at the 0.048 level making the total 0.05. In essence, the researchers decide that the interim results must be highly significant in order to terminate the study.

Basis of the Rule

Repeated examination of a random process must take into account the associated probabilities of Type I and Type II errors.

Discussion and Extensions

As indicated in the rule, multiple comparisons come up in a variety of research settings. A clinical trial usually has endpoints to be shown different under the treatment conditions. These tend to be confirmatory analyses, especially with primary endpoints. Strict adherence to a Type I error is indicated here. This was the basis for the strict allocation of α to the interim analyses in the illustration. On the other hand, many observational studies have no favorite endpoints and are frequently exploratory in nature. It is not uncommon to allow a Type I error of 0.10 or so to make sure that no potentially interesting findings are missed. A third type of multiple comparison procedure comes up in clinical trials with interim testing of results. This is often required for ethical reasons. The usual strategy is to "spend" a little of the Type I error on the interim analyses and reserve most for the final analysis. This is based on an acknowledgment that multiple testing does compromise the Type I error.

The U.S. Food and Drug Administration (FDA) is particularly concerned with the problem of multiple comparisons. In some ways objectives of the FDA differ from that of the pharmaceutical industry in terms of interpretation of clinical trials. The FDA wants to minimize a Type I error (approving an ineffective drug). The pharmaceutical industry wants to minimize a Type II error (failing to get approval for an effective drug). These two aims are statistically incompatible (but both the FDA and the pharmaceutical industry want to protect the public so there is clearly common ground as well). Ignoring the multiple comparisons problem will lead to increased power—clearly very desirable for all parties. But the price is an increased Type I error. In a regulatory context multiple comparisons must be addressed carefully and consistently. There are no formal FDA guidelines (Sankoh et al., 1997) but review committees are very aware of the issue and will take multiple test results into account in their evaluation of safety and efficacy of a new drug application.

The question will be raised which multiple comparison procedure to use. There are two common situations: pairwise comparisons and multiple comparisons. With pairwise comparisons the Tukey procedure is recommended. The Bonferroni method is useful when the data are not balanced. There are technical improvements on Bonferroni such as the Hochberg procedure (see for example, Sankoh et al., 1997) but for a quick rule of thumb the Bonferroni can still be used: Multiply the observed p-values by the number of comparisons made. If the number of comparisons to be made is not known, but apt to be large, the Scheffé method is preferable. It is very conservative but provides complete protection. See Fisher and van Belle (1993) for a discussion of these procedures.

The problem of multiple comparisons continues to bedevil statistical consultants and researchers. Answers range all the way from, "there is no problem" (Rothman, 1990) to rigid adherence to adjustments for multiple comparisons. A very balanced analysis can be found in Proschan and Waclawiw (2000). Schouten (2000) gives an interesting strategy for combining evidence from multiple outcomes in a clinical trial. He suggests defining an outcome measure in the research protocol that summarizes and ranks combined outcome scores from several variables of clinical interest, including side effects.

This chapter has tried to show that purpose, design, conduct, analysis, and interpretation of studies are all tightly linked. Since these form a chain in time, the design is perhaps the most important aspect of an investigation; data from badly designed studies are very difficult to salvage. Next in importance is the conduct of a study. Here the missing data problem raises its head. A study with non-ignorable missing data presents a daunting challenge to the analyst. The child of missing data is often a missed opportunity for a scientific contribution. Analyses and presentations of data are the capstone of the research effort. They should be graceful, to the point, and fair representations of the investigator's efforts. The analyses that are presented are almost certainly a selection from a huge number that have been carried out. Consistency with the original purpose of the study should guide their selection. This way avoids some of the pitfalls of multiple comparisons. Multiple comparisons ironically point back to the beginning—the purpose of the study. If the purpose was clear, the need for multiple comparisons is minimized.

7

Words, Tables, and Graphs

This chapter presents some simple rules of thumb for communicating numerical information in the form of words, tables, or graphs. There are several aspects to numerical information including quantity, variables, and conditions. It can be cross-sectional or longitudinal.

Graphs of data are still a weak point of statistical computing packages. There should be a much tighter link between statistical analysis and graphical output. For example, computer output of an analysis of variance should automatically present graphs of means and standard deviations of the relevant treatment effects. See Figure 6.1 on page 141 for an illustration.

There are excellent books on computer graphics and the visual display of data. Examples are the books by Tufte (1983, 1990, 1997), Cleveland (1985, 1993), and Chambers et al. (1983). One of the classics of graphical statistics is the book by Huff (1954), it is still worth reading and pondering.

7.1 USE TEXT FOR A FEW NUMBERS, TABLES FOR MANY NUMBERS, GRAPHS FOR COMPLEX RELATIONSHIPS

Introduction

Numerical information is primarily displayed in sentences, tables and graphs (Tufte 1983, page 178). When should these be used? Either singly or in combination? The following sections provide some rules of thumb.

Rule of Thumb

Use sentence structure for displaying 2 to 5 numbers, tables for displaying more numerical information, and graphs for complex relationships.

Illustration

Consider the sentence: "The blood type of the population of the United States is approximately 40%, 11%, 4% and 45% A, B, AB, and O, respectively." This is a pretty bad sentence. The reader has to go to the end of the sentence to understand the structure, then go back and forth. Use of "respectively" is usually not a good idea, and not even necessary in this case. The sentence could have been re-written: "The blood type of the population of the United States is approximately 40% A, 11% B, 4% AB, and 45% O." This structure immediately associates blood type with frequency. But a simple sentence table, as follows, is still better:

The blood type of the population of the United States is approximately,

O	45%
A	40%
B	11%
AB	4%.

Note that the blood types have now been ordered, not by alphabet, but by the more meaningful characteristic, frequency. The sentence has been displayed so that the information about blood type is clearly set off.

Basis of the Rule

The content and context of the numerical data determines the most appropriate mode of presentation. A few numbers can be listed, many numbers require a table. Relationships among numbers can be displayed by statistics. However, statistics, of necessity, are summary quantities so they cannot fully display the relationships, so a graph can be used to demonstrate them visually. The attractiveness of the form of the presentation is determined by word layout, data structure, and design.

Discussion and Extensions

Tufte (1983) discusses another reason for constructing a table rather than a graph: when there are many "localized comparisons." Such tables can typically be found as basic descriptors of study populations in clinical trials, epidemiological studies, and social science observational studies. Information in such tables may deal with age, sex, race/ethnicity, and other basic characteristics. It would be difficult, and pointless, to display such information graphically.

A journal editor is not likely to allow authors the luxury of displaying the sentence-table due to space constraints. Readability and design are not the driving criteria for page layout in journals.

If the numerical information is crucial then a graph will not do. Random numbers, decimal expansions of mathematical constants, and values of integrals are always displayed in tables, not graphs. Tufte (1983, page 33) makes the point that "...small, non-comparative, highly labelled data sets usually belong in tables."

Displaying numerical information always involves selection. The process of selection needs to be described so that the reader will not be misled. Even the blood type data is almost certainly based on a non-random selection of the population (blood banks) and the issue of representativeness comes up.

7.2 ARRANGE INFORMATION IN A TABLE TO DRIVE HOME THE MESSAGE

Introduction

There is some flexibility in table structure and content. Structure deals with labeling, content with the values associated with the labels.

Rules of Thumb

- Arrange the rows and columns in a table in a meaningful way in order to display as much structure as possible.

- Limit the number of significant digits.

- Make the table as self-contained as possible.

- Use white space and lines to organize rows and columns.

- Use the table heading to convey crucial information. Do not stint. The more informative the heading, the better the table.

Illustration

Numerical reports from government agencies frequently order the information alphabetically. This is preferable for identifying a particular category such as an occupation, city or state. On the other hand, to display data structure it will be more useful to order the categories in some meaningful way.

Table 7.1, from MacKay et al. (2000), lists the number of active health professionals in the United States according to occupation for 1980. The alphabetical listing of occupations facilitates identifying specific categories such as "Optometrists." This is probably the best arrangement for initial use. But the structure in the data is not very clear. A more meaningful arrangement orders the occupations by their frequency.

Footnotes in the source table indicate that the data were gathered from many agencies. This is clear in the table; the number of speech therapists listed as 50,000 is almost certainly a rounded estimate while the number of physicians, 427,122, is

Table 7.1 Number of Active Health Professionals According to Occupation in 1980: United States[a]

Occupation	1980
Chiropractors	25,600
Dentists	121,240
Nutritionists/Dietiticians	32,000
Nurses, registered	1,272,900
Occupational therapists	25,000
Optometrists	22,330
Pharmacists	142,780
Physical Therapists	50,000
Physicians	427,122
Podiatrists	7,000
Speech therapists	50,000

[a] From National Center for Health Statistics, 2000; Table 104 (Subcategories not included, Only data for 1980 is used). See MacKay et al. (2000).

Table 7.2 Table 7.1 Re-arranged by Number in Occupational Category and Rounded to the Nearest 1000.

Occupation	1980 in 1000's
Nurses, registered	1,273
Physicians	427
Pharmacists	143
Dentists	121
Physical Therapists	50
Speech therapists	50
Nutritionists/Dietiticians	32
Chiropractors	26
Occupational therapists	25
Optometrists	22
Podiatrists	7

most likely the sum of physicians from medical society rolls. This suggests different levels of precision for the entries in the table. In this situation the numbers with lowest precision should determines the number of significant digits: units of 1000's. The edited numbers are displayed in Table 7.2. This table illustrates simple structure among the rows. For example, nurses are the most common health professionals; the next most common are physicians—about one third of the total number of nurses. The table also uses *white space* to group categories with similar frequencies. Judicious use of white space enhances the information in a table.

Table 7.3 extends the data to include active health personnel in 1995-1997 and orders the table by percentage of change and percentage of change adjusting for the growth in population that occurred during the period. This arrangement emphasizes the fastest and slowest changing professions: nutritionists/dieticians and physical therapists versus dentists and pharmacists. The last column in the table also implies that the rate of increase in the number of health professionals was greater than the rate of increase in the population in all categories—suggesting one possible reason for the huge increase in medical costs in the United States.

Basis of the Rule

Columns and rows of tables represent dimensions and must be used creatively and efficiently. Labels may be arbitrary and an ordering based on labels may not display data structure. (Consider what Table 7.1 would have looked like with occupations in Spanish.)

Table 7.3 Number of Active Health Professional in the United States. Table 7.2 Extended to Include Numbers in the Period 1995-1997 and Ordered by Percentage Change. The Last Column Adjusts the Change for Population Size

Occupation	Number in 1980 in 1000's	Number in 1995-1997 in 1000's	Percent Change	Adjusted for Population Growth
Nutritionists/Dietiticians	32	69	116%	85%
Physical Therapists	50	107	114%	84%
Speech therapists	50	97	94%	67%
Chiropractors	26	47	81%	59%
Occupational therapists	25	45	80%	55%
Nurses, registered	1,273	2,162	70%	46%
Physicians	427	724	70%	40%
Podiatrists	7	10	43%	30%
Optometrists	22	30	36%	13%
Dentists	121	160	32%	9%
Pharmacists	143	185	29%	11%

Discussion and Extensions

It is interesting to compare census tables with tables in an almanac. In the latter there is often a ranking: the ten longest rivers in the world, ranking of nations by energy use and so forth. Tables in almanacs tend to display more structure than census tables.

Most tables carry too many digits. For example, the precision of percentages is proportional to the square root of the sample size (see Rule 2.8, page 43). This suggests guidance for percentages as summarized in Table 7.4. If the denominator is based on 10 observations percentages should be rounded to the nearest 10. Only if the denominator is on the order of 100,000 should percentages carry one decimal place. It may seem harsh, but these recommendations are probably too lenient. They assume a Poisson distribution for the numerator—in many cases a conservative assumption since there is extra-Poisson variation.

Continuous measurements have several sources of variation. For example, the age of a person is rounded to years, and age varies from person to person. A reasonable rule of thumb for biological measurements is to start with the assumption of a 30% coefficient of variation. Together with the square root rule for sample size precision this leads to a more sober use of significant digits in tables.

Ehrenberg(1977) uses the term "effective digits," defined as digits that vary in a table. For example, the numbers 354691, 357234, 356991 have four effective digits, not six. Ehrenberg also makes the point that numbers in tables should have at

Table 7.4 Suggested Number of "Significant Digits" for Percentages as a Function of the Size of the Denominator[a]

Denominator	% error	Suggested percentage expression
10	32%	30, 40, 50, 60 etc.
100	10%	35, 40, 45, 50 etc.
1000	3.2%	34, 36, 38, 40 etc.
10,000	1%	34, 35, 36, 37, 38 etc
100,000	0.1%	34.1, 34.2, 34.3, 34.4 etc

[a]Percent error based on Poisson(θ). Coefficient of variation= $100\sqrt{\theta}/\theta = 100/\sqrt{\theta}$. For Example, $32\% = 100/\sqrt{10}$.

most two effective digits. The human mind has difficulty scanning three or more effective digits.

Computer output is notorious for carrying many more digits than are significant. The only justification for this practice is that the output is frequently considered intermediary to further analyses. In that case more digits can, and should, be carried. But these should form the "deep background" of the statistical analysis and should never see the light of day in a publication.

Some judgment must be exercised in using these rules of thumb for tables. For example, if percentages in a table are based on different numerators then a compromise will have to be struck for the sake of uniformity. In the case of percentages based on very small denominators, the reader should be warned that the number of significant digits have to be taken with a grain of salt .

Good references for the construction of tables are Wainer(1997, 1998), his columns in the journal *Chance*, Ehrenberg (1977), and Tufte (1983, 1990, 1997).

Displaying numerical information in a table requires creativity and thought. It is unfair to criticize agencies that produce hundreds of tables routinely for a variety of audiences. Their presentation requires compromises that will affect their overall impact. However, researchers developing a few tables for a publication are not under those constraints. Their tables should reflect an integration of art and science.

7.3 ALWAYS GRAPH THE DATA

Introduction

The virtues of good graphics have already been touted. A good graph illustrates patterns, identifies outliers, and shows relationships that were perhaps unanticipated. There are any number of horror stories associated with ungraphed statistical analyses.

Tufte (1983) provides three criteria for graphical excellence.

1. "Graphical excellence is the well-designed presentation of interesting data—a matter of *substance*, of *statistics*, and of *design*."

2. "Graphical excellence consists of complex ideas communicated with clarity, precision, and efficiency."

3. "Graphical excellence is that which gives to the viewer the greatest number of ideas in the shortest time with the least ink in the smallest space."

Rule of Thumb

When possible always graph the data.

Illustration

One of the most interesting illustrations of the virtues of graphing data was developed by Anscombe (1973) who constructed four sets of regression data with the property that the slopes, intercepts, correlations, and standard errors were identical. This data set has been reproduced many times (for example in Tufte, 1983). Figure 7.1 is a graph of the data. It illustrates that summary numerical information does not adequately describe the relationships among the variables. There are statistics, such as leverage points, that will detect outliers such as occur in the lower right-hand panel of Figure 7.1. However, such statistics were derived based on examination of graphical output. They may not pick up a unique pattern.

Basis of the Rule

A good graph displays relationships and structures that are difficult to detect by merely looking at the data.

Discussion and Extensions

All the comments made in the opening paragraphs of this chapter apply to graphing data. It may be useful to ask when it is not useful to graph data. There are two situations: when there are few data points, or when there are too many relationships to be investigated. In the latter case a table may be more effective.

One way to develop good graphical techniques is to see how the experts do it. The references cited at the beginning of this rule provide a good starting point. Their advice is always worth considering.

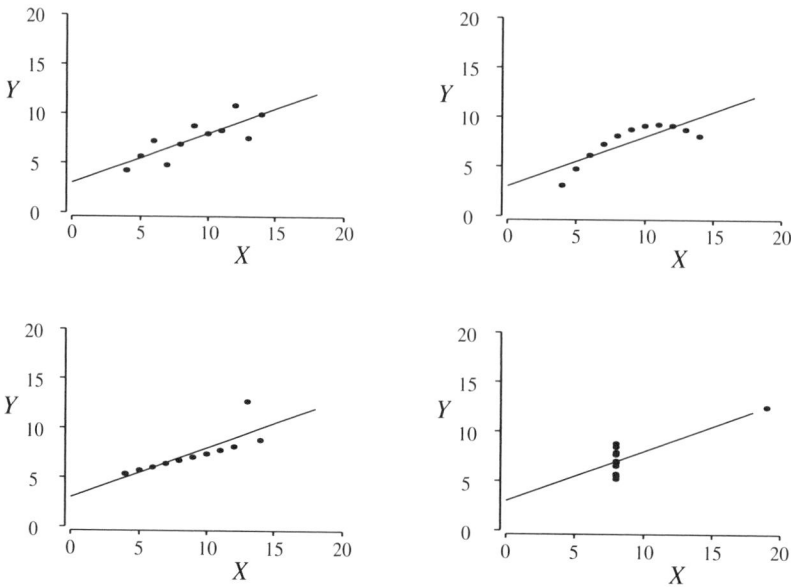

Fig. 7.1   Regression data with the property that for every set the means are equal, the regression lines are $Y = 3 + 0.5X$, the standard errors of estimate of slope are 0.118, and the correlation coefficients are 0.82. Data from Anscombe (1977).

7.4   NEVER USE A PIE CHART

Introduction

The most ubiquitous graph is the pie chart. It is a staple of the business world.

Rule of Thumb

Never use a pie chart. Present a simple list of percentages, or whatever constitutes the divisions of the pie chart.

Illustration

Consider the ABO and Rh systems for the population of the United States as displayed in the pie chart of Figure 7.2. It is clear that the pie chart adds absolutely nothing to the data which is displayed as a simple table in the right-hand panel of the figure. Instead of illumination there is a waste of visual time and effort—with no ultimate reward. Note also that having to go back and forth from the legend to the graph is frustrating, and the hatched lines are tiring. Adding actual percentages to the chart

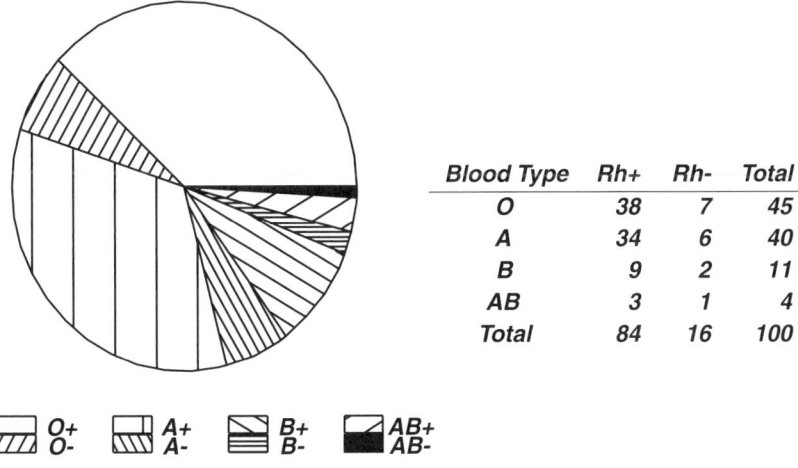

Blood Type	Rh+	Rh-	Total
O	38	7	45
A	34	6	40
B	9	2	11
AB	3	1	4
Total	84	16	100

Fig. 7.2   Pie chart of frequencies of blood type and Rh classification for the population of the United States. Percentages in the right-hand panel form the basis for the pie chart.

would increase the complexity without adding insight. Color could have clarified the relationships somewhat but is expensive to print.

Basis of the Rule

The pie chart has very low data density. In the example above there are eight data points (actually seven since the sum of the percentages adds up to 100). These data points can be presented much better as a table.

Discussion and Extensions

This pie chart violates several of the rules suggested by the question posed in the introduction. First, immediacy: the reader has to turn to the legend to find out what the areas represent; and the lack of color makes it very difficult to determine which area belongs to what code. Second, the underlying structure of the data is completely ignored. Third, a tremendous amount of ink is used to display eight simple numbers. The table in the right panel provides a very simple summary of these data. The table also displays elements of underlying structure indicating, for example, that the OAB and Rh distributions operate independently.

There is a brief, but interesting, discussion of pie charts in Tufte (1983). He cites additional reasons for not using pie charts. He also asserts that the only thing worse than one pie chart is several or many pie charts. There are usually much better ways of presenting such information.

The urge to use a pie chart should be creatively channeled into carefully thinking about the nature of the numerical information, and the purpose of the presentation. If the "take home message" is that the audience remembers two or three numbers, then it is best to present and reinforce the numbers in a variety of ways. If many numbers are involved, the pie chart will be cluttered.

7.5   BARGRAPHS WASTE INK; THEY DON'T ILLUMINATE COMPLEX RELATIONSHIPS

Introduction

Every meeting has presentations that involve bar graphs.

Rule of Thumb

Always think of an alternative to a bar graph.

Illustration

Figure 7.3, Panel A, summarizes smoking status among adults 18 years and over by family income, sex, race, and Hispanic origin in the United States in 1995, from Pamuk et al. (1998). It is difficult to discern differences between men and women and patterns associated with family income. Panel B, provides a better summary of the data. It immediately shows that black and white males have the highest smoking rates, while Hispanic males and females have the lowest smoking rates. At the same time change in smoking frequency with education is highest among the black and white males.

Basis of the Rule

The bar graph is far removed from the original data and there are now better ways of presenting data. Alternatives take more thinking but clearly pay off. Bar graphs are prime examples of "chart junk" defined by Tufte as unnecessary use of ink. The bars take up a great deal of ink, yet the only purpose is to indicate height. Furthermore, a bar graph frequently ignores the underlying structure of the data, as the example indicates.

Discussion and Extensions

A bar graph typically presents either averages or frequencies. It is relatively simple to present raw data (in the form of dot plots or box plots). Such plots provide much more information, and they are closer to the original data. If the bar graph categories are linked in some way—for example, doses of treatments—then a line graph will be much more informative. Very complicated bar graphs containing adjacent bars are very difficult to grasp. If the bar graph represents frequencies, and the abscissa

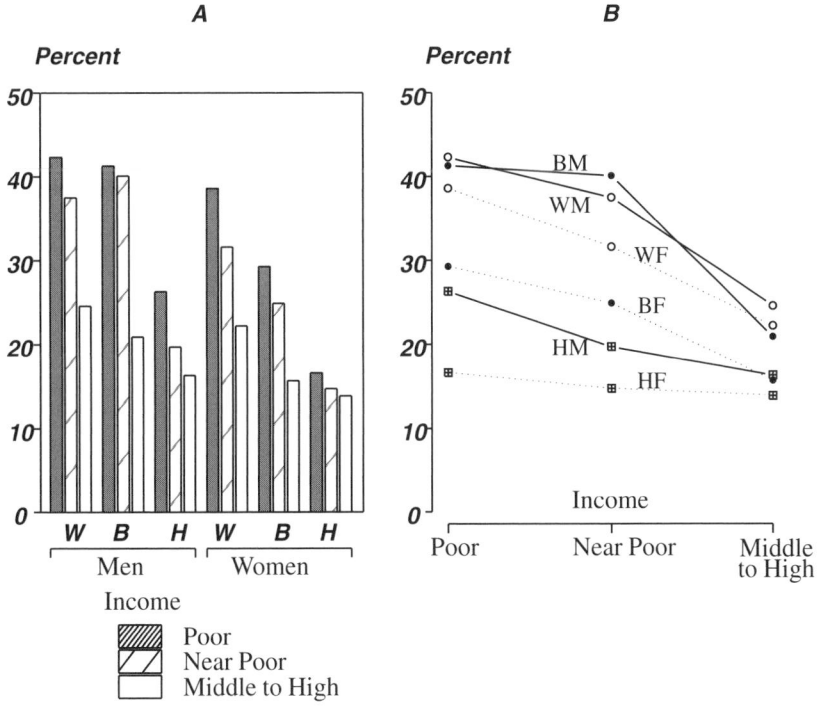

Fig. 7.3 Cigarette smoking (%) in 1995 among adults 18 years of age and over by family income, sex, race, and Hispanic origin for the United States. Adjusted for age. From data for Figure 36, Pamuk et al. (1998). Panel A: original format; Panel B: new format.

values can be ordered, then a line graph will be much more informative and will have substantially reduced chart junk.

Computer packages frequently offer the bar graph as the first graphing option. One way to move beyond this option is to make a free-hand sketch of a possible graph that embodies the points to be made. Then search through the package to select a graph that comes close to the sketch. Unfortunately, many times it will be necessary to construct the graph using a drawing package.

7.6 STACKED BARGRAPHS ARE WORSE THAN BARGRAPHS

Introduction

Stacked bar graphs are seen even more frequently than bar graphs. Their putative purpose is to show structure in the data.

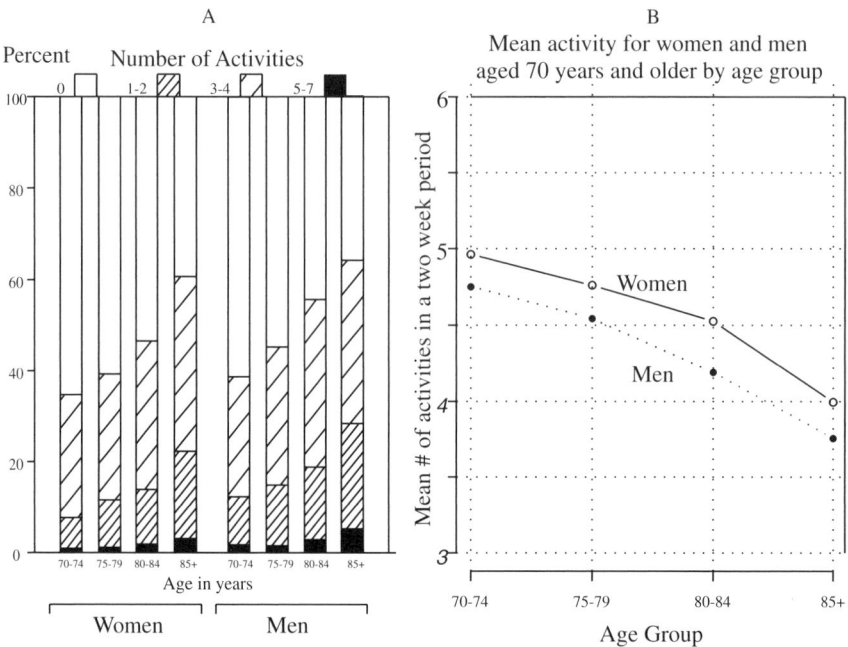

Fig. 7.4   Number of social activities in 1995 in a 2-week period among persons 70 years of age and older. From data for Figure 21 in Kramarov et al. (1999). Panel A: original format; Panel B: new format.

Rule of Thumb

There are much more effective ways of showing data structure than stacked bar graphs.

Illustration

Stacked bar graphs are very difficult to digest. Consider Panel A in Figure 7.4 which deals with the number of social contacts in a two-week period among persons 70 years of age and over, by age, and sex for the United States in 1995 (data from Kramarov et al., 1999). Some of the questions that could be asked:

1. Do men or women have more activities?

2. How do activities vary with age?

3. Do these changes differ between men and women?

It is impossible to answer these questions from the stacked bar graphs. The data for these graphs is presented in Table 7.5. The mean number of activities have been plotted in Panel B of Figure 7.4. The answers to the three questions appear immediately:

Table 7.5 Number of social activities in a two-week period among persons 70 years of age and older[a]

Number of Activities	70-74 years	75-79 years	80-84 years	85 years and over
	%	%	%	%
Women				
0 activities	1.0	1.3	2.1	3.1
1-2 activities	6.8	10.5	11.9	19.2
3-4 activities	26.8	27.5	32.5	38.3
5-7 activities	65.4	60.7	53.5	39.4
Mean number	4.96	4.76	4.53	3.99
Men				
0 activities	1.9	1.7	2.9	5.3
1-2 activities	10.5	13.3	15.9	23.0
3-4 activities	26.3	30.3	36.7	35.9
5-7 activities	61.2	54.7	44.5	35.9
Mean number	4.75	4.54	4.17	3.96

[a] From data for Figure 21 in Kramarov et al. (1999).

1. Women tend to be more socially active than men at all ages.

2. Activities change moderately with age; about 0.5 activities per person over a ten-year age interval.

3. The decline in activities over time appears to be about the same for men and women.

Basis of the Rule

Stacked bar graphs do not show data structure well. A trend in one of the stacked variables has to be deduced by scanning along the vertical bars. This becomes especially difficult when the categories do not move in the same direction.

Discussion and Extensions

The stacked bar graph needs to be analyzed carefully in order to come up with a better way of presenting the data.

It should be noted that the figures in the publications in the above sources are marked by a very complete definition of terms and for every figure there is a table on which the figures are based. The tables also include standard errors which can be used to indicate internal variability. It is obviously desirable to have some measure of variability attached to the graph. This is impossible for the stacked bar graph. Figure 6.1 provides an illustration on how to attach estimates of variability to a graph without cluttering the graph.

These graphs illustrate how to drive home a message about complex relationships among several variables. They also show that there are systematic components to these relationships. A thoughtful graph reinforces verbal descriptions.

7.7 THREE-DIMENSIONAL BARGRAPHS CONSTITUTE MISDIRECTED ARTISTRY

Introduction

Graphs can be embellished in a variety of ways. In Victorian England this might have included zephyrs gently blowing data across the graph. In the 21st century there are even worse embellishments.

Rule of thumb

Never use three-dimensional bargraphs.

Illustration

Figure 7.5 illustrates how this confuses rather than helps the reader. What percentages are associated with the bars? The lower or the upper part?

Basis of the Rule

The embellishment of a third dimension creates confusion.

Discussion and Extensions

It is not clear where the pernicious practice of the third dimension embellishment came from, or why it continues to be so popular. There is absolutely no redeeming feature to this practice.

One of the key characteristics of data is their dimensionality. The dimensions of the data can often be grouped into classes such as outcome variable(s), treatment variable(s), covariate(s), time, and block(s). It makes sense to make the graphs as close to the original data as possible (Diggle et al., 1994). The reason is straightforward: data reduction involves decisions about what is important and what is not important. The reader can participate in this judgment if the original data are presented as closely as possible. Manufacturing additional dimensions obviously violates this recommendation.

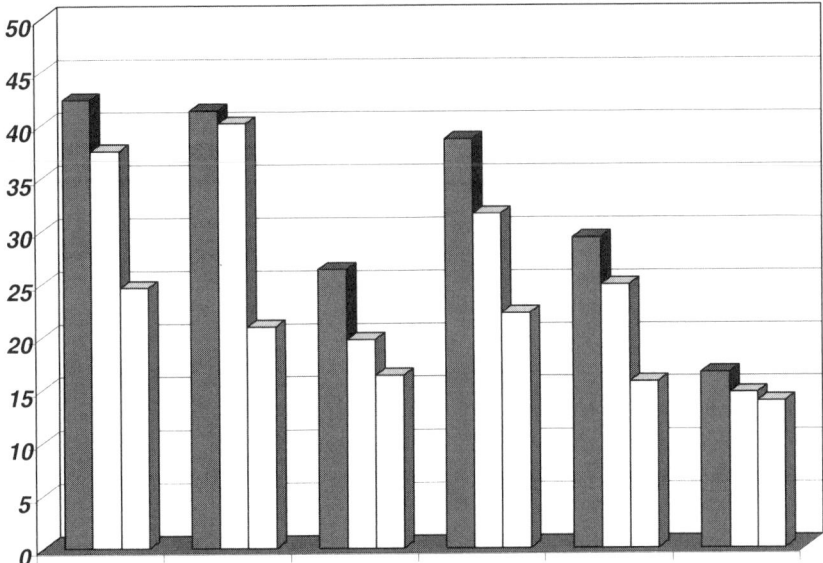

Fig. 7.5 Distortion introduced by adding extra dimension to a bar graph. Same data, and graph, as in Figure 7.3.

To graph an essentially linear variable as an area or a volume does not make much sense. The bar graph presents areas but since the widths are identical the focus is on the heights.

The previous two rules have argued that using bar graphs for displaying frequencies or numbers is not good statistical practice. Before going to a bar graph think of the purpose of the graph and then find a more appropriate way of displaying the relationships.

7.8 IDENTIFY CROSS-SECTIONAL AND LONGITUDINAL PATTERNS IN LONGITUDINAL DATA

Introduction

The last twenty years have seen an explosion of methods for the analysis of longitudinal data, in part due to increased computational resources. The graphical presentation of data presents special challenges because frequently there is both cross-sectional and longitudinal information.

Rule of Thumb

In the case of longitudinal data identify both cross-sectional and longitudinal patterns.

Illustration

Consider the middle panel of Figure 7.6, displaying the population of the United States from 1800 to 2000 for 9 regions. There is no linkage of the regions. The advantage of this panel is that all the data are there. One possible way to show more structure is to use box plots which contain medians, quartiles and minimum and maximum values. These are displayed in the top panel. They give a good idea of the population growth in the past 200 years, and the increasing variation in regional populations. But on the whole, this is not a useful application of box plots since they essentially describe the data cross-sectionally. The bottom panel identifies the regions and links them across time. For example the New England Region comprises Maine, New Hampshire, Vermont, Massachusetts, Rhode Island and Connecticut. The Pacific region consists of the states of Washington, Oregon, California, Alaska and Hawaii. Now it is possible to tell where population growth has taken place in the United States over the last 200 years. The longitudinal analysis indicates the explosive growth of the Pacific and South Atlantic regions: the southern, ocean bordering states. The fan-shaped pattern of frequencies across time provides almost the same cross-sectional information as the box plots, and in addition, shows the relative ranking of the region from one 20 year period to another. A similar analysis by state, rather than region, would show the same kind of pattern.

The second panel in Figure 7.6 could have been constructed by identifying a symbol for each region and with a legend in the margin. This detail would have been informative but visually very difficult to follow.

Basis of the Rule

When there is more than one source of variation it is important to identify those sources.

Discussion and Extensions

Diggle et al. (1994) list four guidelines for graphing longitudinal data. Their advice is useful:

- show as much of the relevant raw data as possible rather than only data summaries;
- highlight aggregate patterns of potential scientific interest;
- identify both cross-sectional and longitudinal patterns;
- make easy the identification of unusual individuals or unusual observations.

The challenge to graphing of longitudinal and cross-sectional data is to display the levels of variability. In Figure 7.6 the variability in any one year is cross-sectional displaying between-region variation. The boxplots in the figure are a good measure of that variability. The slope of a line for a region is a measure of within-region variability. The third panel displays that variability.

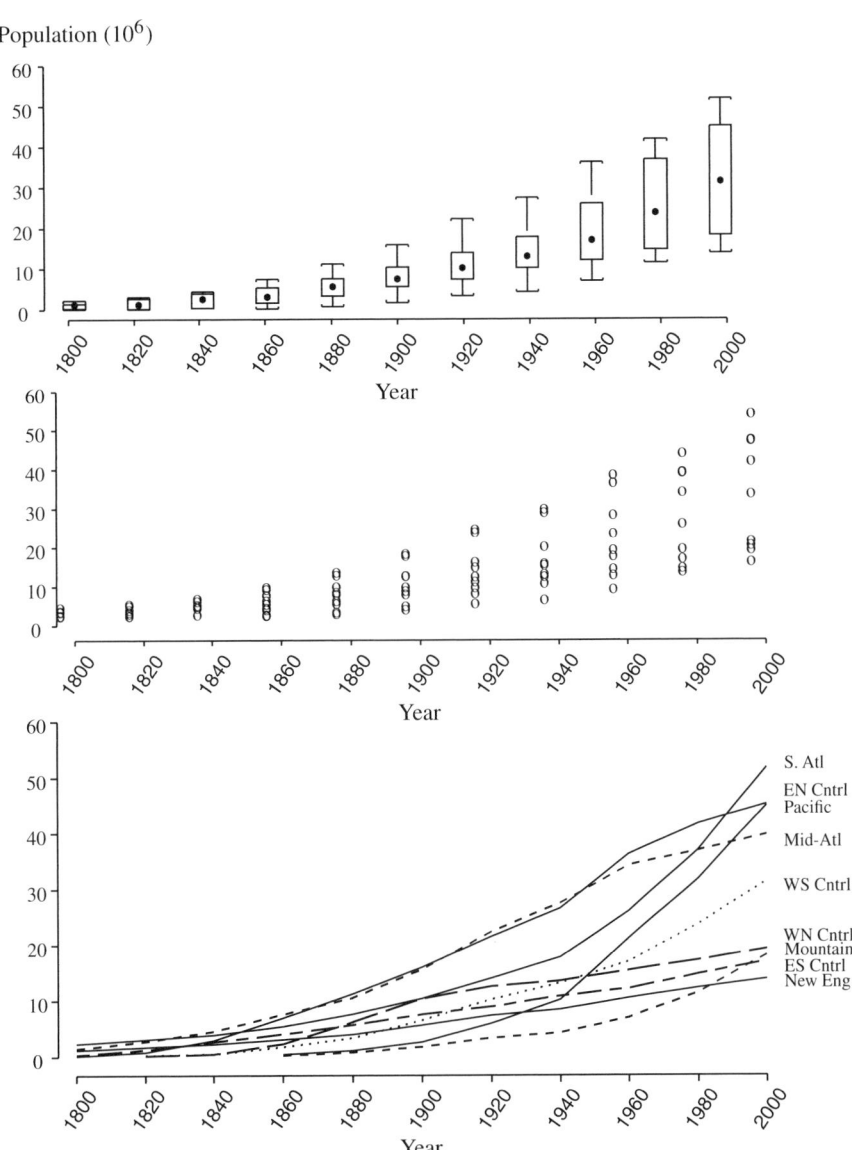

Fig. 7.6   US population by region and year from 1800 to 2000.

A display of longitudinal data can be very informative, indicating patterns that are very difficult to detect or summarize from the numerical values. Formal tests and comparisons can be used to verify the perceived relationships.

7.9   USE RENDERING, MANIPULATION, AND LINKING IN HIGH DIMENSIONAL DATA

Introduction

A great deal of statistical data is high-dimensional—for example, multiple regression data with dozens of predictor variables. Visualizing these kinds of data has intrigued and challenged statisticians for many years. The confluence of high capacity computing, statistical methodological advances, and huge data sets has focused attention on graphical displays of high-dimensional data. Since humans are restricted to visualizing at most three dimensions the challenge is how to approach higher-dimensional data. The ideas in this section are based on the work of Sutherland et al. (2000).

Rule of Thumb

Three key aspects of presenting high dimensional data are: rendering, manipulation, and linking. *Rendering* determines what is to be plotted, *manipulation* determines the structure of the relationships, and *linking* determines what information will be shared between plots or sections of the graph.

Illustration

Figure 7.7 incorporates some of these principles. It presents the estimated % increase in total mortality per $10 \mu g/m^3$ increase in PM_{10} for the 90 largest cities in the United States. The cities have been grouped by region. Within a region cities have been listed by population size.

The widths of the confidence intervals are roughly related to the size of the city. For example, Chicago and Detroit in the Industrial Midwest region have very tight confidence bands. The effect associated with these two cities is small but significantly different from zero. The overall mortality effect for the 90 cities was approximately 0.4% per $10 \mu g/m^3$—a very small effect, but affecting a huge number of people. The figure also indicates that the majority of effects are positive rather than negative.

The computational effort to produce these estimates was huge. Statistical models were specifically developed to model the mortality. See Samet et al. (2000b) for a description.

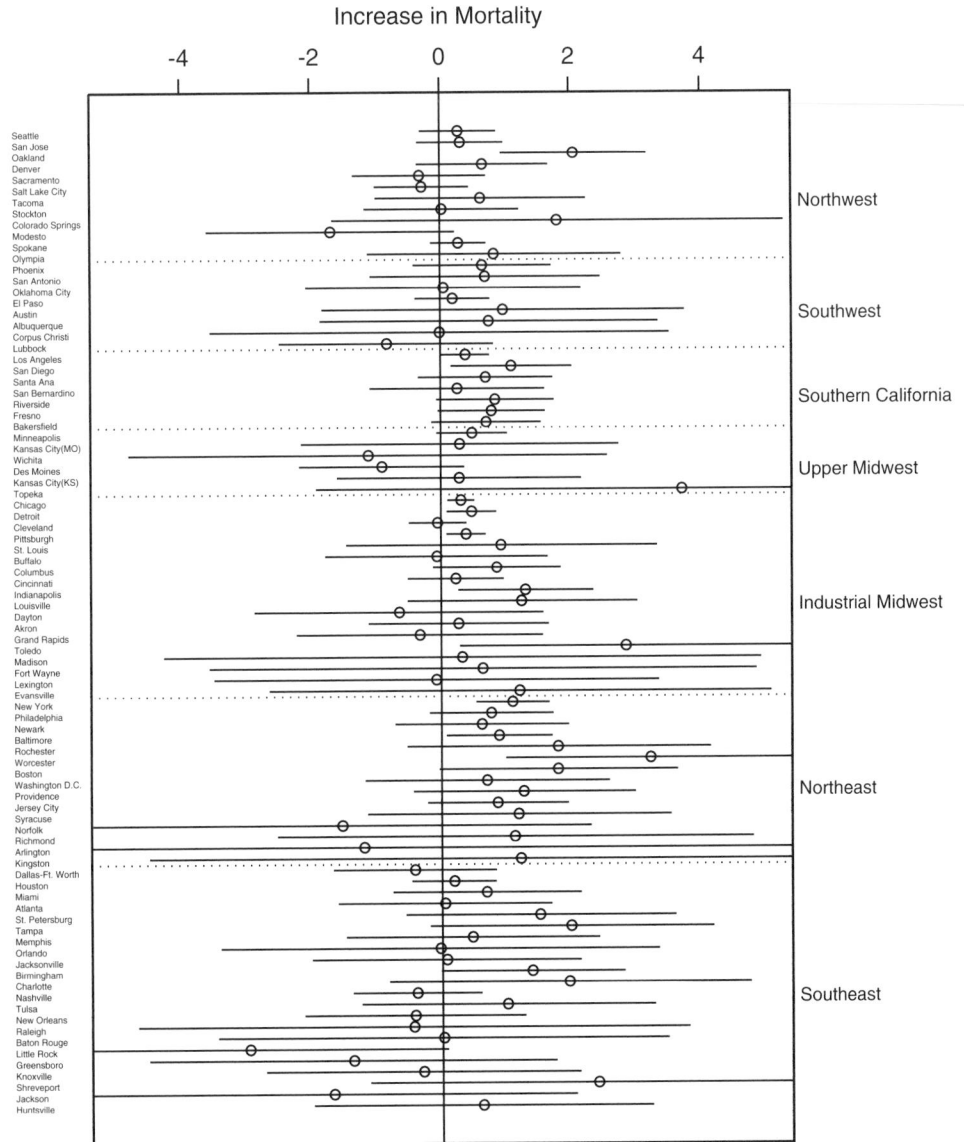

Fig. 7.7 Estimated % increase in total mortality per 10 $\mu g/m^3$ increase in PM_{10} for the 90 largest cities in the United States, grouped by region. Based on Samet et al. (2000b), Data Supplied by F. Dominici.

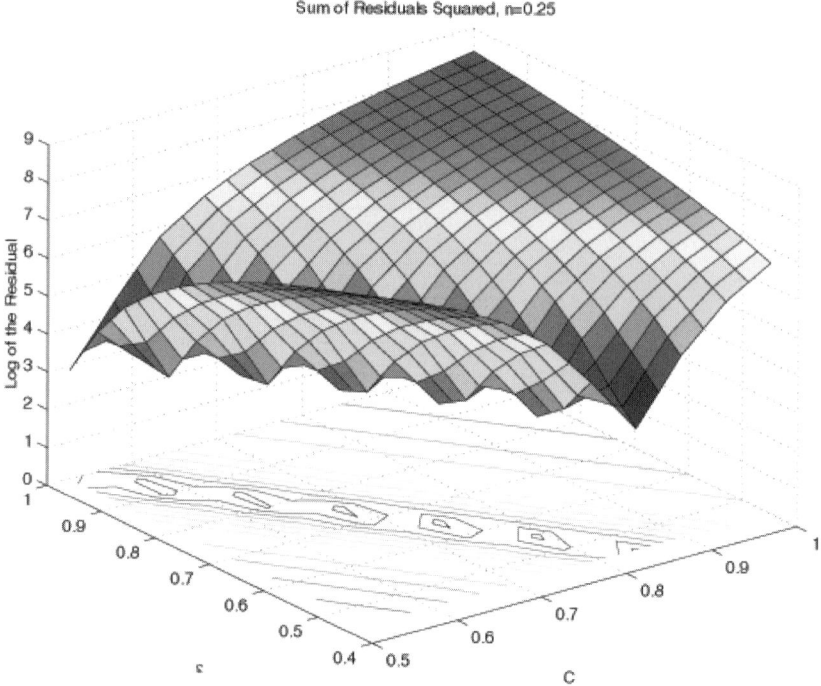

Fig. 7.8 Least squares surface for surface temperatures of a copper rod heated internally. Observations taken every 30 seconds for about twenty minutes. Parameters of interest are the emissivity and a parameter related to convection. Courtesy, Louis van Belle.

A second example shows how high dimensional data can be reduced by looking at residuals from a model; this can be done irrespective of the number of variables or parameters used in the model since the residuals will be unidimensional. The data for Figure 7.8 came from a study investigating heat transfer in a copper rod. The rod was heated internally and a data acquisition system recorded surface temperature at multiple points on the rod every 30 seconds for 20 minutes. The change in temperature was modelled on the basis of a differential equation. Two parameters of interest were emissivity, ϵ, and a parameter, C, related to convection. The least squares surface is shown in Figure 7.8 (rendering). One nice feature of the graph is the isopleths in the X-Y plane indicating that there are many local minima. The linear pattern across the plane also suggests that there is substantial collinearity in the model. This graph could be rotated to look "under" the surface (linking)—some computer packages provide a way of doing this. The precision of the graph depends on the number of grids defined over the region of interest (manipulation). Changing the grid pattern led to a somewhat different graph. This graph illustrates one technique for analyzing high dimensional data: look at the residuals from the model: they form a one-dimensional domain and can be examined under a variety of conditions. In high-dimensional regression models the residuals can be plotted against each of the predictor variables to elucidate potential patterns.

Basis of the Rule

Rendering, manipulation, and linking are basic ways in which relationships can be categorized. The choice of which aspects to render, manipulate, and link are, of course, the hard thing to do. But the rule provides a means of attacking the problem of graphing high-dimensional data.

Discussion and Extensions

Given the high dimensionality of the data structures associated with both figures the choice of which aspects to display is the most challenging. The design of the study should point to which variables will be prominently displayed. It would be misleading to begin a study with emphasis on one set of variables and then, after the data are in, to focus only on another, data-driven, set. This borders on the unethical—especially if the reader is not warned that this was done.

Figure 7.8 is a nice example of good graphics, and the clever use of the three dimensions that are available. Actually, there is a fourth dimension of the variables associated with the contour plots in the X-Y plane. A color version is even more informative.

This chapter provides an organized structure for summarizing numerical information by means words, tables and graphs. The choice between words and tables is primarily a function of the amount of numerical data that has to be described. The choice between a table and a graph is determined by whether a realtionship is to be assessed. This suggests limited or no usefulness for pie charts and bar graphs since they are very poor at displaying relationships. Additional graphs of relationships can be found in the previous chapters.

8

Consulting

The rules enumerated in previous chapters dealt with the substantive aspects of statistical practice. This chapter deals with more process oriented features. The rules to be stated are based on experience and the literature. The terminology will be that of statistical consultant and investigator. The term client will not be used, it has a less collaborative feel to it. It is useful to make a distinction between _cross-sectional consulting_ and _longitudinal consulting_. The former means a brief, somewhat superficial conversation between consultant and investigator. For example, the investigator has received a comment on paper recommending a statistical test of the data and wants to know where to find a statistical package that will carry out the test. Usually, cross-sectional consulting involves statistical advice for data already collected or for analyses already carried out. The investigator will have a specific question or idea about what is wanted (the challenge for the consultant is to find out what is really needed!). Occasionally it may involve planning of studies or experiments. The consultation may take on a didactic tone in that a specific statistical issue is explained to the investigator. Longitudinal statistical consulting, as the name suggests, involves a long-term relationship between a statistician and investigator. Frequently it is collaborative and collegial and may lead to creative application of current statistical methods or the development of new methods. A panel of the Institute of Mathematical Statistics called these two types _advisory_ and _interactive consulting_ (IMS Panel, 1990). It is not the case that longitudinal consulting is inherently more satisfying, or more valuable, than cross-sectional consulting. The choice of one over the other may depend on the work environment.

The literature on statistical consulting is substantial. It ranges from personal reminiscences of senior statisticians recounting their life experiences as consultants (see for example, Finney, 1982), to didactic articles and discussions (see, for example,

Chatfield, 1991, and Hand, 1994), to books on consulting such as Boen and Zahn (1982) and Rustagi and Wolfe (1982). A special mention is Cox (1999), who states in about two pages what others have used articles and books for; see Rule 8.12 on page 190. These are first-class resources. Many other articles deal with experiences of cross-sectional consulting, typically listing horror stories and the statistical response. While instructive, there is a sameness about them: The stories are interesting but somehow miss the excitement of statistical consultation, occasionally lapsing into a condescending tone. They miss the ambiguity of a great deal of data analysis, the ethical binds, the pressures of time, the genuine and valid excitement of the scientists, and the satisfaction of a stimulating collaborative enterprise.

8.1 STRUCTURE A CONSULTATION SESSION TO HAVE A BEGINNING, A MIDDLE, AND AN END

Introduction

Rather than start with what constitutes good statistical consulting, it is less daunting to describe a good statistical consulting session. However, even distilling the essence of a good statistical consulting session is not a simple task. Many criteria can be constructed. Here is a rule that seems central.

Rule of Thumb

Structure a consulting session to have beginning, a middle, and an end. The session begins with problem context and definition, moves to resolution and solution, and concludes with a summary and allocation of responsibilities.

Illustration

A colleague asks for "help with a meta-analysis of the pediatric literature on the use of ritalin." This is a first-time visit although the consultant knows the researcher. The investigator should introduce the context and content of the problem (beginning). The consultant should review the assumptions and process of meta-analysis–keeping in mind that this may not be the issue that ultimately should be addressed—it may well be decided that a meta-analysis is not what is needed and alternative considerations may be discussed. The session ends with a discussion of next steps, allocation of responsibilities, and summary.

Basis of the Rule

Consultation involves process and substance. It is important that the consultant be aware of the process and be able to implement it. Structuring the consulting session with a clear steps and product in view will pace and guide the process.

Discussion and Extensions

This rule is one of the most difficult to implement. A consulting session frequently is scheduled for a fixed length of time such as an hour. Ideally, the colleague has provided a short write-up of the topic to be discussed, or has sent some papers that deal with the topic. That will allow structuring the discussion more effectively, shortening and sharpening the beginning of the consultation. The consultant needs to judge what can and cannot be accomplished in that time. The challenge is that problem definition and context may take more time than planned. This can be due to the difficulty of the topic, the loquaciousness of the investigator (or the consultant!), or the realization that a Type III error is about to be committed: the right answer to the wrong question. The possibility of the Type III error means that the good consultant figures out what the question should be, not just what the question is.

As soon as both parties agree that the pace is going to be slower than anticipated, a "time out" should be called to settle on what will be accomplished during the session.

Good statistical practice includes writing a brief, one page summary of the session for the investigator and for the files. This will, typically, include description of the subject area and problem, statistical issues, decisions and recommendations, and action items. Such a file is useful for closing the consultation loop; and documenting consulting activity which can then be summarized easily for administrative purposes.

8.2 ASK QUESTIONS

Introduction

One way to learn is by asking questions. A good question implies that the questioner is beginning to understand the content area, has a sense of where the discussion is going, and wants to make sure that every one agrees on the direction.

Rule of Thumb

Pace the consultation by asking focused, helpful questions.

Illustration

If the research involves Alzheimer's disease, ask questions about the diagnosis (by exclusion), what diseases have to be ruled out (multi-infarct dementia, Parkinsonism), how progression is measured (by cognitive tests and activities of daily living), and so on.

Basis of the Rule

Asking focused questions shows active listening, paces the conversation, allows the investigator to reflect, and may turn up surprising answers.

Discussion and Extensions

All commentaries on statistical consulting mention the key role of asking questions. The challenge is to ask focused questions. It is usually easy to ask general questions. Some people develop a skill of asking unanswerable questions, which seem to be intended to show the superiority of the questioner. Consultees are usually very good at picking up this motivation and will not react positively. Questions need always be asked in a spirit of humility and desire for knowledge.

It is sometimes difficult to know where to begin with questions. The following rule provides one approach with a checklist of topics to think about in the consultation.

8.3   MAKE DISTINCTIONS

Introduction

How do consultants learn? Active listening (or reading) and questioning are certainly beginnings. But soon the consultant is faced with a mountain of information. Then what should be done?

Rule of Thumb

Make valid distinctions, not distinctions valid.

Illustration

If an investigator presents a large number of undifferentiated endpoints, ask which ones are the key ones. That is, help the investigator distinguish among the endpoints. This can be done by asking which endpoint will be emphasized in an abstract. If a grant proposal is discussed, the key endpoint will most likely be the one on which sample size calculations will be based. Another way to get at this is to ask about fundamental results in the research area. All these are intended to help the investigator develop an ordering of endpoints. Focusing on key endpoints will help the investigator come to a clearer understanding of numerical aspects of the investigation.

Basis of the Rule

Making appropriate distinctions is one of the hallmarks of constructive critical thinking.

Discussion and Extensions

At an initial consultation it is particularly important to distinguish what is central from what is peripheral. To illustrate, the investigator may ask for interpretation of

Table 8.1 Basic Distinctions to be Made in a Consulting Session

Item	Choice I	Choice II
Intent of study	Exploratory	Confirmatory
Source of data	Random	Convenience
Sampling unit	Simple	Multiple
Randomization	Single	Multiple
Model	Fixed	Random or mixed
Type of study (1)	Intrinsic	Comparative
Type of study (2)	Cross-sectional	Longitudinal
Epidemiologic study	Cohort	Case-Control
Categories	Nominal	Some kind of ordering
Numerator/denominator	Rate	Proportion, ratio

statistical output. If a regression analysis using SAS was carried out, the consultant should be able to distinguish between Type I, II, and III analyses, and be able to explain the rationale for choosing one over the other. The appropriateness of one analysis over another will involve discussion with the investigator about the selection of the data and the purpose of the analysis. Table 8.1 lists some basic distinctions consultants should be able to make.

A useful distinction can be made among types of studies. A broad classification is *intrinsic studies, comparative studies* and *equivalence studies*. An intrinsic study estimates a quantity on its own merits—for example, the proportion of AIDS patients in a population or the urinary excretion rate of males over the age of 65. Another type of intrinsic study deals with the assessment of association between two or more variables. A comparative study, as the name suggests, wants to make a comparison between two or more quantities. Most clinical trials are comparative with the comparison of new treatments or therapies with standards or placebo. Another example is to compare the radon concentration in homes with and without basement fans. The equivalence study is a special type of comparative study with the purpose of demonstrating equivalence of therapies, measurement techniques, or associations. For example, the comparison of generic and brand-name drugs deals with equivalence. These distinctions affect design, sample size, and analysis. Hence, the consultant should determine fairly early in the consultation what kind of study is being contemplated.

Distinctions have been emphasized in this book. Chapter 3 discussed measures of association and distinctions that had to be made among them.

Another crucial distinction is between fixed, random, and mixed effect models. This may come up more frequently in longitudinal consulting. It is important that the consultant know which analysis is appropriate for the data.

Knowing when to make a distinction and when to accept the current status—and perhaps ambiguity—is part statistical skill, part wisdom.

8.4 KNOW YOURSELF, KNOW THE INVESTIGATOR

Introduction

It is impossible to know everything. The very fact that there is a consultation implies that two parties (at least) want to learn and exchange information. This exchange, ideally, should be mutually beneficial. The maximum benefit is obtained if it is very clear what each party brings to the table.

Rule of Thumb

Know yourself, and know the investigator.

Illustration

Good consultants know what they know and don't know. Although they are being consulted, they do not feel obligated to have immediate answers to all questions. If they don't have the answer, they say so and provide a road map to the answer. This may involve doing some research outside the consulting session, asking a colleague for advice, or concluding that the problem is not answerable. Investigators react more to the attitude about a possible lack of knowledge than to the lack of knowledge.

Basis of the Rule

Investigators have a way of probing the limits of the consultant's knowledge. In fact, they frequently come because they were not able to apply the standard examples found in many textbooks.

Discussion and Extensions.

Knowledge of self is more difficult than it appears. It requires standing outside of oneself and looking in. Part of this knowledge is gained in graduate school by (hopefully) sympathetic instructors. After that it may take a friend to do an inventory of strengths and weaknesses. A knowledgeable, sympathetic mentor is an unusual privilege.

Consultants should be aware of how they think. It has been asserted that statisticians tend to think abstractly whereas researchers tend to think concretely (Hall, 2000). These are just that—tendencies—but it is important to be aware of ways of approaching a problem. Some people move from the general to the particular, and others tend to move in the opposite direction. While it is not necessary to do a personality inventory before starting a consulting session, it is important to know how a research issue is approached and to be sensitive to the investigator's mode of thinking.

One of the more difficult professional challenges is working as the sole statistician; in this situation it is especially important to develop resources for help when the answers to the problem are not apparent.

Knowing the investigator means that the consultant has some understanding of the investigator's background knowledge of statistics. If this is not known before the consulting session, then some time needs to be spent on this topic during the consulting session. Learning how to probe in a constructive manner is a required consulting skill.

Good consultants, and good investigators, are very aware of the limits of their knowledge. This is not a matter of false humility, rather it is a realistic judgment about limits and the need to move beyond them. People who are very clear about this usually are very good and very quick learners.

8.5   TAILOR ADVICE TO THE LEVEL OF THE INVESTIGATOR

Introduction

The previous section raises a general issue about communicating with investigators. This is particularly challenging in cross-sectional consulting. A large part of an initial consulting session will be spent in learning terms and vocabulary. The initial burden is on the statistician: In order to become conversant with the investigator's problem, the statistician must dig into the field. This means asking questions, summarizing, and restating what the investigator said. But at some time it will become necessary for the consultant to introduce concepts that may be foreign to the investigator.

Rule of Thumb

Tailor advice to the investigator's level of understanding. Avoid jargon, equations, and formulas.

Illustration

If the investigator knows about paired t-tests but has never heard of randomized blocks—which may be useful for the proposed investigation, begin with explaining how the paired t-test is a special case of a randomized block and, perhaps, how the t-test is a special case of the F-test.

Basis of the Rule

The meaning of the word "communication" is a "process by which information is exchanged between individuals through a common system of symbols, signs, or behavior." The stress in statistical consultation is on the common system of symbols, signs, or behavior. In fact, all three are independent contributors to good—or bad—consultation.

One strategy for tailoring advice is to keep things as simple as possible, but not more simple (another application of Rule 1.5, page 11). Symbols, equations, and formulas are the natural domain of statistics and statisticians but they may put the investigator to sleep or induce the MEGO syndrome: My Eyes Glaze Over.

When consulting for the first time there is a tendency for graduate students to see an application of the latest methodology they just heard about (very much like the medical student who "sees" the disease discussed in the last medical lecture). It is often the case that a simple analysis will validly answer most (if not all) of the investigator's questions.

One place where tailoring is especially important is in a grant application. Most applications require a "statistical section." And frequently the statistician will be asked to write that section (perhaps a day before the grant application is due). Such a section will require all the statistician's skills. It will need to be tailored to the research area, the research proposal—and the reviewers! If sample size calculations are needed this is where they will go; these may have to be generic if there are no explicit null and alternative hypotheses. A generic sample size calculation postulates effects in terms of standardized differences thus obviating the need for specific values for effects and variability. This approach is not recommended since savvy reviewers will immediately sense a lack of statistical involvement in the grant application.

Long-term consulting builds mutual understanding of skills and knowledge. In this situation, the consultant will learn and use the appropriate level of advice for the colleague.

8.6   USE UNITS THE INVESTIGATOR IS COMFORTABLE WITH

Introduction

The symbols for sensitivity, specificity, and prevalence in Chapter 4 were made deliberately evocative in order to speed the comprehension of the idea. This is particularly important in discussions with nonstatisticians. The following rule is suggested.

Rule of Thumb

When communicating with nonstatisticians, use natural units as much as possible.

Table 8.2 Natural Units (Counts) for Assessing Screening Effectiveness for Colorectal Cancer by Means of the Hemoccult Test

	Disease Status		
Test	Colorectal Ca	No Colorectal Ca	Total
Hem test pos	150	2,991	3,141
Hem test neg	150	96,709	96,859
Total	300	99,700	100,000

Illustration

Consider the example in Chapter 4 dealing with screening studies (Rule 4.10, page 96). The hemoccult test for colorectal cancer had sensitivity of 50%, specificity of 97%, and a prevalence of disease of 0.003. The predictive value of a positive test was 5%. This was obtained using Bayes' theorem. However, it could also have been derived by generating Table 8.2 describing a population of 100,000 people. The entries in the table follow directly from the given statistics.   This table contains all the information about sensitivity, specificity, prevalence, and predictive value of positive and negative tests. For example, the predictive values of a positive is seen immediately to be 150/3,141 = 4.8%. The prevalence is 0.003 and so on. The table also shows that out of 100,000 people screened, only 3,141 can be expected to be positive on the test. This has profound public health implications. From the table, costs associated with each test can immediately be translated into total public health costs. These numbers can also be scaled up to, say, 1,000,000 if a larger screening population is envisioned.

Basis of the Rule

It has been said that probability theory is hard—even for probabilists. This is doubly true for nonprobabilists. A "natural" unit is not easy to define explicitly, but it basically refers to a valid unit closest to the investigator's experience and comfort zone.

Discussion and Extensions

The example was taken from Hoffrage et al. (2000), who quote a study in which one group of physicians (n = 24) was presented the data as probabilities and another group (n = 24) was presented with frequencies as given in Table 8.2. Both groups were asked to calculate the predictive value of a positive test. In the first group, 1/24 gave the correct answer; in the second group, 16/24 gave the correct answer.

A symbolic statement of Bayes' theorem is more generalizable than a table of frequencies, but there are alternative ways to help people generalize from known results. The concepts of prevalence, sensitivity, and specificity are intuitive when applied to Table 8.2 and provide all the information needed to generalize to other tables.

Using units the consultee is familiar with is one more way in which the statistician can tailor advice to achieve the best possible communication.

8.7 AGREE ON ASSIGNMENT OF RESPONSIBILITIES

Introduction

Statistical consultation as a collegial activity generates expectations and responsibilities.

Rule of Thumb

When the consulting session is done, be clear as to who does what, and when.

Illustration

In a consulting session, data analysis may come up. Make clear who is going to enter the data, verify data entry, decide what statistical package will be used, and determine who will run the analyses.

Basis of the Rule

Consulting is like raising children. Everyone is happier if the rules and limits are clear.

Discussion and Extensions

One example of being clear about the product of a consultation is potential authorship or co-authorship on a paper. Be fairly clear about this. One advantage of longitudinal consulting is that this kind of issue will (should?) have been thrashed out and consultation can begin without having to worry about this. This is likely to be more of an issue in an academic environment.

Another academic issue is use of the data in a methodology paper, perhaps to illustrate a statistical method. If none of the data have yet appeared in print, the owner is less likely to grant permission to use the data—particularly if the owner is a "slow" publisher.

A related issue is acknowledgments. What if the statistician is acknowledged in a paper on the basis of a hallway conversation about statistical methods, the advice clearly was not taken, and the analysis is, in fact, dubiously correct? At this point very little can be done. Acknowledgments are usually seen as an expression of appreciation by the author of the paper. In the case of statistical advice, less appreciation would have been more welcome. The consultant can ask not to be acknowledged unless the paper has been reviewed beforehand. The Department of Biostatistics at the University of Washington has a policy of not allowing acknowledgments. But this is difficult to enforce without appearing pedantic and condescending.

8.8   ANY BASIC STATISTICAL COMPUTING PACKAGE WILL DO

Introduction

Next to questions about sample size, the consultant is most often asked two additional questions: "What is a good statistical package?" and "What is a good book?"

Rule of Thumb

Any basic statistical package will do, any text book will do.

Illustration

A t-test is a t-test is a t-test in Minitab, DataDesk, Stata, SPSS, SAS, Statgraphics, Statview, GraphPad,....

Basis of the Rule

By this time most of the standard statistical packages have been tested and are reliable.

Discussion and Extensions

Chances are that investigators have organized their data using Excel. Statisticians tend to look down on this use of a spreadsheet for statistical analysis. But its almost universal use indicates that it meets a universal need. Part of its attractiveness is its capability for data storage, data analysis, and data graphing all in the same package. All this can be transferred to text-writing programs almost seamlessly. The statistician may be used to a particular package but should not force the investigator to use it. For example, many statisticians use S+ or SAS, but their investigators may use a spreadsheet package or a more basic package such as Minitab or Statview. It is counterproductive to stress the virtues of one package over another at an initial consultation.

With respect to statistical texts, similar advice applies. Most basic texts are accurate. But investigators will wander from one text to another in the hope of finding an answer to a problem they may not really understand. Typically they will scour one text after another for an example that fits their situation. If the investigator wants to work at that level of understanding, find the text that contains that example, make a copy, and go on to other issues.

8.9 ETHICS PRECEDES, GUIDES, AND FOLLOWS CONSULTATION

Introduction

Ethics is often what the other person should follow. Ethics deals with the right and good conduct of consultation. Some ethical principles are fairly straightforward: Don't lie, don't fudge or cook data, don't deliberately slant the report of a study. There are situations that are more challenging—areas of research that conflict with the statistician's value system. Two examples: research involving animals and randomized placebo–controlled clinical trials.

It is less clear whether or not to take part in an analysis of data that were obtained unethically, such as valuable data obtained without informed consent, older data obtained under less stringent ethical guidelines, and data from countries with very different ethical standards.

Rule of Thumb

Ethics precedes, guides, and follows consultation.

Illustration

If the statistician does not want to be involved in research with animals, the potential investigator should be informed as soon as possible with the suggestion of the name of a colleague who would be willing to help. This should be communicated in a facilitative manner, not defensively or aggressively.

Basis of the Rule

Ethics shapes the context of a consultation. While many aspects of the context are understood or part of the ethos of the scientific community, there are basic values that may have to be enunciated before collegial consulting can take place.

Discussion and Extensions

Many statistical groups have deliberated on a code of professional ethics for statisticians. For example, the International Statistical Institute (1985) has a very detailed statement about ethics. There is also a later statement by the American Statistical Association (1999). The ISI statement talks about *obligations* to society, funders and employers, colleagues and subjects.The ASA talks about *responsibilities* to funders, clients, employers, research subjects, colleagues, other statisticians, and statistical practitioners. It also discusses responsibilities regarding allegations of misconduct, and concludes with the other side of the coin: responsibilities of employers and organizations employing statistical practitioners. Both statements can be found on these organization's web sites. The ISI statement emphasizes internal, personal ethical obligations, whereas the ASA statement has a flavor of externally imposed responsibilities. These are very much two sides of the same coin.

The consultant may be obligated to withdraw from a study for a variety of reasons. It may be that the investigator does not want to follow the proffered advice, or the investigator asks for advice at the end when the study may not salvageable. In each case the consultation process should be structured in such a way so that the consultant can withdraw gracefully and promptly.

If there is outright fraud, or grossly unethical manipulation of data is suspected, the consultant needs to be aware of a process for dealing with the issue—for example, a confidential discussion with the consultant's supervisor. The supervisor may have to take actions that the consultant did not anticipate, such as starting a formal investigation. The consultant should be clear about the consequences of any actions in this kind of situation.

While outright fraud is rare, there are grayer areas of unethical data acquisition or manipulation that may violate personal ethical standards. Again, self-knowledge is the key to deciding what is acceptable and what is not.

Ethical standards for scientific investigations have changed over time and vary from country to country. This does not mean that "it's all relative." In fact, the opposite is true. Collegial reflections implicitly assume that there are standards, that they can be discussed, that they can be described, and that they can be achieved. Such standards not only influence the aims of an investigation but also the process (dwelling exclusively on process is a sign of relativism).

8.10 BE PROACTIVE IN STATISTICAL CONSULTING

Introduction

Statisticians in many ways are the methodologists of the scientific method. This creates obligations, responsibilities, and opportunities.

Rule of Thumb

Be proactive. Being proactive means taking the lead in scientific discussions, coordinating separate research efforts, and thinking broadly about the scientific enterprise.

Illustration

The example comes from work with the Alzheimer Disease Research Center at the University of Washington. This is a geographically and scientifically dispersed group spread over many locations and representing research expertise in neurology, neuropathology, neuropsychology, molecular biology, and statistical genetics. As part of the biostatistical core, other cores and projects were invited to visit and describe their research. This educated the statistical members of the center, gave the researchers an opportunity to share their excitement, and allowed the biostatistics core to comment on the projects in a nonthreatening way.

Basis of the Rule

Statistical consulting is best done at the start of research or production projects. By being proactive the statistician indicates commitment to such projects.

Discussion and Extensions

While the rule is simple to state and is clearly correct, it is more difficult to describe how to implement it. This depends very much on the work situation. Basically, the statistician needs to be perceived as approachable, knowledgeable, and willing to learn. For example, one excellent way to establish rapport is to attend seminars in the research area. The goal is to become knowledgeable enough to be able to ask appropriate questions. Ask colleagues to suggest key journals in their field and then read them carefully. Attend national meetings of the subject matter area.

In a collaborative research environment there may be a central database environment managed by the statistician. In this kind of situation the statistician has an overview of the total research effort that individual collaborators may not be aware of. This creates an opportunity for linking the collaborative efforts. It also can be demanding in that each investigator assumes that the data base is exclusive for their use and requests for analysis will be filled instantly.

In committee meetings, the designated statistician must not only be seen but also heard. Being heard implies having something to contribute and, therefore, obligates thorough understanding of the subject matter area. This also means that the statistician must be prepared and must have studied the agenda. Frequently a key issue will deal with the statistical validity of the proposed design and the analyses. It will be necessary to make a judgment about their appropriateness and to explain why, or why not. Pacing is also crucial. At times it will pay to start the discussion by stating some questions that should be answered during the discussion. If the questions are perceptive, outlining valid areas of concern, the committee will appreciate this proactive stance. If the questions are not central, don't start the discussion off; wait until others have had their say.

Being proactive does not imply perfection. G.K. Chesterton said, "If a thing is worth doing, it is worth doing badly." What he meant was that it is better to try and succeed in part than to keep postponing until perfection is reached. A few halting words spoken in another language in another country are much more appreciated than no attempt at all. The consultant's willingness to become knowledgeable in a particular research area will be much appreciated.

8.11   USE THE WEB FOR REFERENCE, RESOURCE, AND EDUCATION

Introduction

Use of the WEB (short version of World Wide Web) has exploded, and the end is not in sight. As always, the technology drives and enables a great deal of methodology. For example, in the field of clinical trials it is now possible in multi-center clinical trials to enter patients remotely and get treatment assignments 24 hours per day. The WEB is altering database technology, data analysis, and hence statistical practice. A problem with the WEB is that there is no quality control or cautionary advice about statistical procedures.

Rule of Thumb

Use the WEB for reference, resource, and education—for consultants and investigators. But use it with discernment.

Illustration

Using the search engine Google and typing in "sample size" the first three entries were:
```
http://www.surveysystem.com/sscalc.htm
http://www.stat.uiowa.edu/~rlenth/Power/
http://www.researchinfo.com/docs/calculators/index.cfm
```
All of these are excellent; the second WEB site, maintained by Russ Lenth of the University of Iowa is an outstanding resource.

Basis of the Rule

Somewhere on the WEB is a reliable, authoritative answer to most of the basic statistical consulting questions. The trick is to find it, evaluate it, and apply it.

Discussion and Extensions

Start with a good search engine such as Google. The above example illustrates the power of this search engine. After a few searches the same URL's (Universal Resource Locator) will appear and a pattern will emerge. Add these to the library of useful URLs. The list will grow very long very quickly. Another problem is that some sites disappear while others appear so that the URL used last week may no longer work this week. Paradoxically, the search will also turn up sites that are very old (say, 10 years or more) and that have not been updated and are no longer useful. These are the ghost towns of the Internet world. Here are five URL's that are worth looking into. They deal primarily with sample size but will also point to other sites such as statistical packages.

1. `http://www-sci.lib.uci.edu/HSG/RefCalculators.html`
   This is a monstrously large site with every conceivable calculator listed. A fun site to browse.

2. `http://www.math.yorku.ca/SCS/StatResource.html`
   This site, at York University, Toronto, Canada, specializes in ANOVA sample size calculations. There are links to other sites.

3. `http://ebook.stat.ucla.edu/calculators/powercalc/`
   A very good calculator for most two sample size situations.

4. `http://hedwig.mgh.harvard.edu/size.html`
   This calculator deals with epidemiological applications.

A question may come up on how to reference a material on a WEB (in addition to specifying the URL). The following is an example illustration

Kock, N., Avison, D., Baskerville, R., Myers, M., and Wood-Harper, T. (Eds.) (1999), IS Action Research: Can We Serve Two Masters?, International Conference on Information Systems Internet Panel Supplement: `http://ww2.cis.temple.edu/kock/ICIS99/ISARpanel/`. (accessed 1/21/2002)

There is a downside to the WEB just as there is with statistical packages: Statistical analysis untouched by the human mind. The professional statistician may not be in danger of this, but investigators almost certainly will be: Sample sizes may be estimated using the wrong statistical model, on-line analyses will not take into account the lack of randomization of the data, estimates may be based on independent observations when they are in fact paired, and so on. Do not assume that a WEB-based analysis brought by the investigator will be the appropriate one. Ask questions.

8.12 LISTEN TO, AND HEED THE ADVICE OF EXPERTS IN THE FIELD

Introduction

Consulting is a communal activity. It pays to heed the advice of experts in the field.

Rule of Thumb

Listen to, and heed the advice of experts in the field.

Illustration

Sir David Cox has provided wise advice about consulting in *Liaison*, the publication of the Statistical Society of Canada. Many of the rules discussed in this book are applications of his advice which could have been inserted at the appropriate places in the book. For completeness, they are presented here (reprinted with permission of the author and *Liaison*).

1. *If possible, collaborate (i.e., work with an investigator over a period of time) rather than consult (i.e., some occasional discussion of very specific statistical issues with the investigator). Be interested in the subject matter involved. Aim to use terminology of the subject matter field where it differs from common statistical usage. If, on reflection, the investigator seems misguided, retreat from the consultation as soon as politeness and practicality permit. If collaborating, go to the subject matter seminars from time to time, and read the journals in the field. Discreetly determine how much understanding of statistical issues the investigator has. Mechanical use of significance tests to confirm overwhelming effects, for example, is a bad sign.*

2. *Frequently review what is being done to check that the statistical analysis addresses the correct questions. This may help the investigator clarify thinking as well as protect against the most common error in statistical work—answering the wrong question.*

3. *Aim, if feasible, to see some raw data, to understand the measurement processes involved, and to have some appreciation of the general quality of the data.*

4. *Enquire into aspects of the study design that might have bearing on the appropriate analysis.*

5. *Begin with very simple methods.*

6. *If possible, end with simple methods.*

7. *Since nice ideas for analysis often do not work the first time, be prepared to do some modification.*

8. *Do not be frightened to make strong assumptions. When a preliminary answer has been obtained, then consider which of the assumptions may be crucial.*

9. *Take considerable care over presentation and conclusion.*

10. *If your work is to be acknowledged in a paper or a report, ask firmly to see what is written before it is submitted.*

11. *If you feel you should have been a co-author and have not been invited to be, pause for a few days. If, on reflection, you still feel the same, speak quietly to the friendliest of the investigators pointing out, assuming it is true, that you have spent a lot of time and thought on the work.*

12. *Occasionally, very rarely one hopes, be prepared to say that the data are incapable of throwing useful light on the issues involved.*

13. *Find a good balance between thinking things out for yourself and obtaining advice from statistical colleagues (and, of course, therefore, finding time to help them in return).*

14. *If more than 10 percent of what you do ends up by being directly useful, you are doing well.*

15. *If the investigator begins by saying he has a trivial little problem which he is sure you will be able to sort out immediately, don't altogether believe him.*

Basis of the Rule

"That it is unwise to be heedless ourselves while we are giving advice to others, I will show in a few lines"—Ovid.

Discussion and Extensions

The above guidelines give the quintessence of good statistical collaboration. Each one of them is worth pondering and applying. For example, combining the suggestions of starting simple and making strong assumptions are the bases for most of the discussions in the first five chapters: Sample size calculations are based on the normal distributions, epidemiological rules are based primarily on the Poisson and exponential distributions, and environmental rules are based on the lognormal distribution. Each of these distributions are approximations to real-life situations; good statistical consulting knows when to back off, or introduce modifications. But it starts with them.

Statistical methods have become the core methodology of many of the sciences. Consulting questions, consequently, may come from a very diverse group of investigators. Part of the fun of statistical consultation is this breadth of problems and personalities. Another exciting aspect about consulting is the wide applicability of statistical methods. For example, queuing theory methods apply to epidemiology (incidence, prevalence, and duration of disease) and environmental studies (time series analyses of effects of air pollution on health). This requires unusual breadth of statistical knowledge, and continual updating of knowledge in methodology and applications.

Epilogue

This book has presented some statistical rules of thumb for application to a wide variety of disciplines. Each rule has been illustrated and the basis for the rule is given. At times the basis is the application of a theorem, as in the formula for sample size calculation (equation (2.3) on page 31); at other times a more general basis is provided (Rule 1.5 on page 11). These justifications indicate that mathematics is only one pillar that supports the statistical framework. Other pillars are broad philosophical considerations, and issues peculiar to a particular subject area.

The rules in Chapter 1 deal with the basic ingredients of statistics. Some of them address issues sometimes forgotten in the excitement of running a statistical analysis or applying a new technique from the latest statistical package. The rules are cautionary tales.

Chapter 2 is a "bread and butter" chapter starting with a simple sample size formula and widening its applications and implications. Other chapters deal with sample size issues as well but the basic steps are developed here.

Chapter 3 introduces the concept of covariation. A large portion of the scientific enterprise can be couched in terms of questions of covariation. The strongest form is covariation as causation. The weakest is empirically observed association without any kind of mechanism envisioned.

Epidemiology, discussed in Chapter 4, has become a very quantitative area. Epidemiologists, dealing primarily with observational data, have thought carefully about making valid inferences. Inferences deal with rates (prevalence, incidence, hazard) and comparison of these rates (relative risk, odds ratio, hazard ratio, risk difference). There are many threats to validity including, an incorrect statistical model, insufficient sample size, selection and information bias, variable and incomplete follow-up, missing data, censoring, confounding, effect modification, and competing risks. Some of the solutions to these problems include developing an appropriate statistical model, matching and covariate adjustment techniques, imputation, and sensitivity analysis.

Statisticians have always played a role in developing and analyzing environmental data. Chapter 5 presents some rules that have wider applicability but are particularly pertinent to this area. The study of risk has been stimulated and framed by environmental issues.

Chapter 6 could have been the starting chapter for this book. It is the practical implementation and further explication of several of the rules discussed in Chapter 1. Randomization, in all of its forms, continues to provide the gold standard for valid statistical inferences.

Chapter 7 addresses the challenging question of how to present numerical data. Will it be in words, in tables or in graphs? If words, how shall it be said? If tables, how will they be formed? If graphs, what are some good ways of presenting data? The chapter contains several "thou shalt nots." This is actually quite liberating since the prohibitions are few, and the opportunities remaining are virtually infinite. There are additional graphical examples in other earlier chapters.

Chapter 8 provides consulting statisticians with some rules of thumb for consulting. They will benefit in many ways by considering the advice of the experts cited in this chapter. In the spirit of full disclosure, researchers will also find useful rules here. In fact, reading this chapter will allow researchers to make better use of the skills of their statistical colleagues.

The final rule for the applied statistician? Be involved and understand the research areas that form the arena for statistical practice. It provides exciting opportunities for applying the statistics learned, and continuing to be learned.

The final rule for the researcher? Understand and be involved in the statistical analysis. There is enjoyment in thinking quantitatively about a research area. Quantitative thinking at its best represents valid reductions of a very complicated world, a summary of the numerical aspects of reality.

What these rules have shown is that a little algebra can lead to surprisingly useful insights into the design and analysis of both observational and experimental studies.

This concludes a very personal tour of statistical rules of thumb—not rules of thump; they are suggestions for conducting research. It is one statistician's tour of applied statistics. Most of the rules have face validity but there will be quarrels with their generality and applicability. These will be considered sins of commission. In addition, the reader will have rules that have been omitted and should have been included. These are sins of omission. In both cases, comments can be sent to http:\\www.vanbelle.org for recording and suggestions.

References

Adams, M.M. (1987). *William Ockham*, Volume I. Notre Dame University Press, Notre Dame, IN.

Agresti, A. (1990). *Categorical Data Analysis*. John Wiley & Sons, New York.

Ahnn, S. and Anderson, S.J. (1998). Sample size determination in complex clinical trials comparing more than two groups for survival endpoints. *Statistics in Medicine*, **17**: 2525–2534.

American Statistical Association (1999). Ethical guideline for statistical practice. http://www.amstat.org/profession/ethicalstatistics.html.

Anscombe, F.J. (1973). Graphs in statistical analysis. *The American Statistician*, **27**: 17-21.

Armitage, P. and Colton, T. (editors) (1998). *Encyclopedia of Biostatistics*. John Wiley and Sons, Chicester, UK.

Arnold, A. (1992). *Nonparametric Estimation of Reliability*. Ph.D. Dissertation, University of Washington, Seattle, WA.

Atkinson, A.C. and Cox, D.R. (1988). Transformations. In *Encyclopedia of Statistical Sciences*, Volume 9, Kotz, S., Johnson, N.L. and Read, C.B. (editors). John Wiley and Sons, New York, pp.312–318.

Bailar, J.C., Louis, T., Lavori, P.W. and Polansky, M. (1984). A classification for medical research reports. *New England Journal of Medicine*, **311**: 1482–1487.

Barcikowski, R.S. and Stevens, J. (1975). A Monte Carlo study of the stability of canonical correlations, canonical weights and canonical variate-variable correlations. *Multivariate Behavioral Research*, **10**: 353–364.

Barnett, V. and O'Hagan, A. (1997). *Setting Environmental Standards: The Statistical Approach to Handling Uncertainty and Variation.* Chapman & Hall, London.

Bate, R. (editor) (1997). *What Risk?* Butterworth-Heineman, Oxford.

Benson, K. and Hartz, A.J. (2000). A comparison of observational studies and randomized, controlled trials. *New England Journal of Medicine*, **342**: 1876–1886. Discussion in (2000) **342**: 1907–1909 and **343**: 1194–1197.

Bland, J.M. and Altman, D.G. (1995). Comparing methods of measurement: why plotting difference against standard method is misleading. *The Lancet*, **346**: 1085–1087.

Boen, J.R. and Zahn, D.A. (1982). *The Human Side of Statistical Consulting.* Wadsworth, Belmont, CA.

Bowen, W.M. and Bennett, C.A. (editors) (1988). *Statistical Methods for Nuclear Materials Management.* NUREG/CR–4604; PNL-5849. Pacific Northwest Laboratory, Richland, WA.

Box, G.E.P. (1953). Non-normality and tests on variances. *Biometrika*, **40**: 318–335.

Box, G. E. P. (1954). Some theorems on quadratic forms applied in the study of analysis of variance problems. *Annals of Mathematical Statistics*, **25**: 290–302 484–498.

Box, G.E.P. and Jenkins, G.M. (1976). *Time Series Analysis: Forecasting and Control.* Prentice Hall, Englewood Cliffs, NJ.

Boyko, E.J. (1994). Ruling out or ruling in disease with the most sensitive or specific diagnostic test: short cut or wrong turn? *Medical Decision Making*, **14**: 175–179.

Breslow, N.E. and Clayton, D. (1993). Approximate inference in generalized linear mixed models. *Journal of the American Statistical Association*, **88**: 9–25.

Breslow, N.E. and Day, N.E. (1987). *Statistical Methods in Cancer Research.* volume II. International Agency for Research on Cancer Scientific Publications No. 82, Lyon.

Bryk, A.S. and Raudenbush, S.W. (1992). *Hierarchical Linear Models: Applications and Data Analysis Methods.* Sage, London.

Chambers, J.M., Cleveland, W.S., Kleiner, B. and Tukey, P.A. (1983). *Graphical Methods for Data Analysis.* Wadsworth, Belmont, CA.

Chambers, R.L. (1997). Discussion of the paper by Copas and Li. *Journal of the Royal Statistical Society*, **59**: 81.

Chatfield, C. (1991). Avoiding statistical pitfalls (with discussion). *Statistical Sciences*, **6**: 240–268.

Cleveland, W.S. (1985). *The Elements of Graphing Data.* Wadsworth, Monterey, CA.

Cleveland, W.S. (1993). *Visualizing Data.* Hobart, Summit, NJ.

Clayton, D. and Hills, M. (1993). *Statistical Models in Epidemiology.* Oxford University Press, Oxford.

Cochran, W.G. (1947). Some consequences when the assumptions for the analysis of variance are not satisfied. *Biometrics*, **3**: 22–38.

Cochrane, A.L., St. Leger, A.S. and Moore, F. (1978a). Health service 'input' and mortality 'output' in developed countries. *Journal of Epidemiology and Community Health*, **32**: 200–205.

Cochrane, A.L., St. Leger, A.S. and Moore, F. (1978b). The anomaly that wouldn't go away. *The Lancet*, **2:** 1153.

Cohen, J. (1983). The cost of dichotomization. *Applied Psychological Measurement*, **7:** 249–253.

Collett, D. (1998). Sample size determination in survival analysis. In *Encyclopedia of Biostatistics*, Volume 5. Armitage, P. and Colton, T. (editors). John Wiley and Sons, Chichester, U.K., pp.3910–3914.

Concato, J., Shah, N. and Horwitz, R.I. (2000). Randomized, controlled trials, observational studies, and the hierarchy of research design. *New England Journal of Medicine*, **342:** 1887–1892. Discussion in (2000) **342:** 1907–1909 and **343:** 1194.

Cook, R.J. and Farewell, V.T. (1995). Conditional inference for subject-specific and marginal agreement: Two families of agreement measures. (Correction 1998, **26:** 391). *Canadian Journal of Statistics,* **23:** 333–344.

Copas, J.B. and Li, H.G. (1997). Inference for non-random samples (with discussion). *Journal of the Royal Statistical Society*, **59:** 55–95.

Copas, J.B. and Shi, J.Q. (2000). Reanalysis of epidemiological evidence on lung cancer and passive smoking. *British Medical Journal*, **320:**417–418. Responses in *British Medical Journal*, **321:**1221–1222.

Cox, L.H., Guttorp, P., Sampson, P.D., Caccia, D. and Thompson, M.L. (1999). Preliminary statistical examination of the effects of uncertainty and variability on environmental regulatory standards for ozone. In *Environmental Statistics: Analyzing Data for Environmental Policy*, V. Barnett (editor). John Wiley and Sons, Chichester, U.K., pp.122–143.

Cox, D.R. (1999). Some remarks on consulting. *Liaison* (Statistical Society of Canada) **13:** 28–30.

Cox, D.R. and Reid, N. (2000). *The Theory of the Design of Experiments*. Chapman & Hall/CRC, London.

Crouch, E.A.C. and Wilson, R. (1982). *Risk Benefit Analysis*. Ballinger Publishing Company, Cambridge, MA.

Cullen, A.C. and Frey, H.C. (1999). *Probabilistic Techniques in Exposure Assessment*. Plenum Press, New York.

Cullen, A.C., Guttorp, P. and Smith, R.L. (2000). Editorial. *Environmetrics*, **11:** 609-610.

David, H.A. (1985). Bias of s^2 under dependence. *The American Statistician*. **39:** 201.

Davison, A.C. and Hinkley, D.V. (1997). *Bootstrap Methods and Their Application*, Cambridge University Press, Cambridge.

Diggle, P.J., Liang, K-Y., and Zeger, S.L. (1994). *Analysis of Longitudinal Data*. Oxford Science Publications, Oxford.

Doll, R. (1955). Etiology of lung cancer. *Advances in Cancer Research,* **3:** 1–50.

Dorfman, R. (1943). The detection of defective members of a large population. *Annals of Mathematical Statistics*, **14:** 436–440.

Draper, N. and Smith, H. (1998). *Applied Regression Analysis*, third edition. John Wiley and Sons. New York.

Echt, D.S., Leibson, P.R., Mitchell, L.B., Peters, R.W., Obias-Manno, D., Barker, A.H., Arensberg, D., Baker, A., Friedman, H.L et al. (1991). Mortality and morbidity in patients receiving ecainide, flecainide, or placebo. The Cardiac Arrhythmia Suppression Trial. *New England Journal of Medicine*, **324**: 782–788.

Efron, B. (1979). Bootstrap methods: Another look at the jackknife. *The Annals of Statistics*, **7**: 1–26.

Efron, B. (2002). The bootstrap and modern statistics. In Raftery, A.E., Tanner, M.A. and Wells, M.T. (editors). *Statistics in the 21st Century*. Chapman & Hall/CRC, Boca Raton, FL.

Ehrenberg, A.S.C. (1977). Rudiments of numeracy. *Journal of the Royal Statistical Society*, Series A, **140**: 277–297.

Embretson, S.E. and Reise, S.P. (2000). *Item Response Theory for Psychologists*. Lawrence Earlbaum Associates, Mahwah, NJ.

epiTRENDS (2001). Survey reveals variability in emergency care for heart disease and stroke in Washington State. Washington State Department of Health, Olympia, WA. November.

Finney, D.J. (1978). *Statistical Methods in Biological Assays,* 3rd ed. Griffin, London.

Finney, D.J. (1982). The questioning statistician. *Statistics in Medicine,* **1**: 5–13.

Fisher, R.A. (1935). *The Design of Experiments*. Oliver and Boyd, Edinburgh.

Fisher, R.A. (1958). *Statistical Methods for Research Workers,* thirteenth edition—revised. Oliver and Boyd, London.

Fisher, L.D. and van Belle, G. (1993). *Biostatistics: A Methodology for the Health Sciences*. John Wiley and Sons, New York.

Fleiss, J.L. (1981). *Statistical Methods for Rates and Proportions*, second edition. John Wiley and Sons, New York, N.Y.

Ford, E.D. (2001). *Scientific Method for Ecological Research*. Cambridge University Press, Cambridge.

Fraser, D.A.S. (1976). *Probability and Statistics: Theory and Applications*, Duxbury Press, North Scituate, MA.

Friedman, L.M., Furberg, C.D. and DeMets, D.L. (1998). *Fundamentals of Clinical Trials*, third edition. Springer, New York.

Gastwirth, J.L. and Rubin, H. (1971). Effect of dependence on the level of some one-sample tests. *Journal of the American Statistical Association,* **66**: 816–820.

Gehan, E.A. (1980). The training of statisticians for cooperative clinical trials: working statistician's viewpoint. *Biometrics*, **36**: 699–706.

Gelman, A. (2000). Should we take measurements at an intermediate design point? *Biostatistics*, **1**: 27–34.

Glass, G.V., Peckham, P.D. and Sanders, J.R. (1972). Consequences of failure to meet the assumptions underlying the fixed effects analysis of variance and covariance. *Reviews in Educational Research,* **42**: 237–288.

Guttorp, P. (1995). *Stochastic Modeling of Scientific Data*. Chapman & Hall, London.

Guttorp, P. (2000). Setting environmental standards: a statistician's perspective. Technical Report Number 48. National Research Center for Statistics and the Environment, University of Washington, Seattle, WA.

Halfhill, T.R. (1995). Intel's P6. *Byte*. April, 42–58.

Hall, J.W. (2000). The transition from student to consultant. *Liaison* (Statistical Society of Canada). **14:** 24-27.

Hand, D.J. (1994). Deconstructing statistical questions (with discussion). *Journal of the Royal Statistical Society,* Series A, **157:** 317–356.

Hanley, J.A. and Lippman-Hand, A. (1983). If nothing goes wrong, is everything alright? *Journal of the American Statistical Association,* **249:** 1743–1745.

Harvey, A.C. (1981). *Time Series Models.* John Wiley and Sons, New York.

Haseman, J.K. (1978). Exact sample sizes for the use with the Fisher-Irwin test for 2 × 2 tables. *Biometrics,* **34:** 106–109.

Hayes, B. (2001). Randomness as a resource. *American Scientist,* **89:** 300–304.

Hoffrage, U., Lindsay, S., Hertwig, R. and Gigerenzer, G. (2000). Communicating statistical information. *Science,* **290:** 2261.

Hollander, M. and Wolfe, D.A. (1999). *Nonparametric Statistical Methods,* second edition. John Wiley and Sons, New York.

Horton, N.J. and Lipsitz, S.R. (2001). Multiple imputation in practice: comparison of software packages for regression models with missing variables. *The American Statistician,* **55:** 244–254.

Hotelling, H. (1961). The behavior of some standard statistical tests under nonstandard conditions. *Proceedings of the Fourth Berkeley Symposium,* **1:** 319–359.

Hsieh, F.Y. (1989). Sample size tables for logistic regression. *Statistics in Medicine,* **8:** 795–802.

Huff, D. (1954). *How To Lie With Statistics.* W.W. Norton, New York.

Hurlbert, S.H. (1984). Pseudoreplication and the design of ecological field experiments. *Ecological Monographs,* **54:** 198–211.

Husted, J.A., Cook, R.J., Farewell, V.T. and Gladman, D.D. (2000). Methods for assessing responsiveness: a critical review and recommendations. *Journal of Clinical Epidemiology,* **53:** 459–468.

IMS Panel on Cross-Disciplinary Research in the Statistical Sciences (1990). Cross-disciplinary research in the statistical sciences. *Statistical Sciences,* **5:** 121–146.

International Statistical Institute (1985). Declaration on Professional Ethics. http://www.cbs.nl/isi/ethics.htm

Johnson, R.A. and Wichern, D.W. (1998). *Applied Multivariate Analysis.* fourth edition. Prentice-Hall, New York.

Kendall, M.G., Stuart, A. and Ord, J.K. (1987). *Kendall's Advanced Theory of Statistics. Volume 1: Distribution Theory*, fifth edition. Oxford University Press, New York.

Kerlikowske, K., Grady, D., Barclay, J., Sickles, E.A. and Ernster, V. (1996). Likelihood ratios for modern screening mammography. Risk of breast cancer based on age and mammographic interpretation. *Journal of the American Medical Association,* **276:** 39-43.

Kleinbaum, D.G., Kupper, L.L., Muller, K.E. and Nizam, A. (1998). *Applied Regression Analysis and Other Multivariable Methods.* third edition. Brooks/Cole, Pacific Grove, CA.

Kramarov, E., Lentzner, H., Rooks, R., Weeks, J. and Saydah, S. (1999). *Health and Aging Chartbook. Health, United States, 1999* National Center for Health Statistics, Hyattsville, MD. ('http://www.cdc.gov/nchs/data/hus/hus99.pdf)

Kronmal, R.A. (1993). Spurious correlation and the fallacy of the ratio standard revisited (with discussion). *Journal Royal Statistical Society,* Series A, **156:** 379–392.

Kruskal, W.H. and Mosteller, F. (1979a). Representative sampling. I. Scientific literature. *International Statistical Review,* **47:** 13–24.

Kruskal, W.H. and Mosteller, F. (1979b). Representative sampling. II. Scientific literature, excluding statistics. *International Statistical Review,* **47:** 111–128.

Kruskal, W.H. and Mosteller, F. (1979c). Representative sampling. III. Scientific literature, current statistical literature. *International Statistical Review,* **47:** 245–265.

Kruskal, W.H. and Mosteller, F. (1980). Representative sampling. IV. The history of the concept in statistics, 1895–1939. *International Statistical Review,* **48:** 169–195.

Lachin, J.M. (1998). Sample size determination. In *Encyclopedia of Biostatistics,* Volume 5, Armitage, P. and Colton, T. (editors). John Wiley and Sons, Chichester, U.K., pp.3892–3903.

Lachin, J.M. (2000). *Biostatistical Methods.* John Wiley and Sons, New York.

Lakatos, E. (1998). Sample size for clinical trials. In *Encyclopedia of Biostatistics,* Volume 5, Armitage, P. and Colton, T. (editors). John Wiley and Sons, Chichester, U.K., pp.3903–3910.

Land, C.E. (1972). An evaluation of approximate confidence interval estimation methods for lognormal means. *Technometrics,* **14:** 145–158.

Leisenring, W. and Pepe, M.S. (1998). Regression modelling of diagnostic likelihood ratios for the evaluation of medical care. *Biometrics,* **54:** 444-452.

Lettenmaier, D.P. (1976). Detection of trends in water quality data from data with dependent observations. *Water Resources Research,* **12:** 1037–1046.

Lewis, E.J., Hunsicker, L.G., Lan, S., Rohde, R.D., Lachin, J.M. and the Lupus Nephritis Study Group. (1992). A controlled trial of plasmapheresis therapy in severe lupus nephritis. *New England Journal of Medicine,* **326:** 1373-1379.

Liang, K-Y. and Zeger, S.L. (1986). Longitudinal data analysis using generalized linear models. *Biometrika,* **73:** 13–22.

Lin, L.I. (1989). A concordance correlation coefficient to evaluate reproducibility. *Biometrics,* **45:** 255–268.

Lissitz, R.W. and Chardos, S. (1975). A study of the effect of the violation of the assumption of independent sampling upon the type I error rate of the two group t-test. *Educational and Psychological Measurement,* **35:** 353–359.

Little, R.J.A. (1998). Missing data. In *Encyclopedia of Biostatistics,* Volume 4, Armitage, P. and Colton, T. (editors). John Wiley and Sons, Chichester, U.K., pp.2622–2635.

Little, R.J.A. and Rubin, D.B. (1987). *Statistical Analysis with Missing Data.* John Wiley and Sons, New York.

Loh, W-Y. (1987). Does the correlation coefficient really measure the degree of clustering around a line? *Journal of Educational Statistics*, **12:** 235-239.

Looney, S.W. and Gulledge, T.R., Jr. (1985). Use of the correlation coefficient with normal probability plots. (Comments 1995, **39:** 236). *The American Statistician*, **39:** 75-79.

Lord, F.M. and Novick, M.R. (1968). *Statistical Theories of Mental Test Scores.* Addison-Wesley, Reading, MA.

Louis, T. (1981). Confidence intervals for a binomial parameter after observing no successes. *The American Statistician*, **35:** 154.

Lumley, T. and Sheppard, L. (2000). Assessing seasonal confounding and model selection bias in air pollution epidemiology using positive and negative control analyses. *Environmetrics*, **11:** 705–719.

MacKay, A.P., Fingerhut, L.A. and Duran, C.R. (2000). *Adolescent Health Chartbook.* Health, United States, 2000. National Center for Health Statistics, Hyattsville, MD.

Mantel, N. (1951). Rapid estimation of standard errors of the mean of small samples. *The American Statistician*, **5:** 26–27.

Marascuilo, L.A. and McSweeney, M. (1977). *Nonparametric and Distribution-Free Methods for the Social Sciences.* Brooks/Cole, Monterey, CA.

March, R.H. (1996). *Physics for Poets*, fourth edition. McGraw-Hill, New York.

McMichael, A.J., Anderson, H.R., Brunekreef, B. and Cohen, A.J. (1998). Inappropriate use of daily mortality analyses to estimate longer-term mortality effects of air pollution. *International Journal of Epidemiology*, **27:** 450–453.

Meier, P., Free, S.M., Jr. and Jackson, G.L. (1958). Reconsideration of methodology in studies of pain relief. *Biometrics*, **14:** 330–342.

Metz, C.E. (1978). Basic principles of ROC analysis. *Seminars in Nuclear Medicine*, **4:** 283–298.

Millard, S.P. (1987a). Environmental monitoring, statistics, and the law: Room for improvement. *The American Statistician*, **41:,** 249–259.

Millard, S.P. (1987b). Proof of safety vs proof of hazard. *Biometrics*, **43:**719–725.

Millard, S.P. and Neerchal, N.K. (2001). *Environmental Statistics*. CRC Press, Boca Raton, FL.

Millard, S.P., Yearsley, J.R. and Lettenmaier, D.P. (1985). Space–time correlation and its effect on methods for detecting aquatic ecological change. *Canadian Journal of Fisheries and Aquatic Science*, **42:** 1391–1400. Correction: (1986), **43:** 1680.

Mode, N.A., Conquest, L. and Marker, D.A. (1999). Ranked set sampling for ecological research: Accounting for the total costs of sampling. *Environmetrics*, **10:** 179–194.

Murray, G.D. (1998). Missing data in clinical trials. In *Encyclopedia of Biostatistics*,Volume 4, Armitage, P. and Colton, T. (editors). John Wiley and Sons, Chichester, U.K., pp.2637–2641.

Murray, C.J. and Nelson, C.R. (2000). State-space modeling of the relationship between air quality and mortality. *Journal of the Air and Waste Management Association*, **50:** 1075–1080.

National Research Council (1983). *Risk Assessment in the Federal Government*. A Report of the Committee on the Institutional Means for Assessment of Risks in Public Health. National Academy Press, Washington, DC.

Nelder, J.A. and Wedderbrun, R.W.M. (1972). Generalized linear models. *Journal of the Royal Statistical Society,* Series A, **135:** 370–384.

Neville, A.M. and Holder, R.L. (1995). Letter to the editor. *Journal of the Royal Statistical Society,* Series A, **158:** 619–625.

Newman, S.E. (2001). *Biostatistical Methods in Epidemiology*. John Wiley and Sons, New York.

Neyman, J. (1952). *Lectures and Conferences on Mathematical Statistics*, second edition. United States Department of Agriculture, Washington, DC.

Oakes, M. (1990). *Statistical Inference*. Epidemiology Resources Inc., Chestnut Hill, MA.

Ott, W.R. (1995a). *Environmental Statistics and Data Analysis*. Lewis Publishers, Boca Raton, FL.

Ott, W.R. (1995b). Human exposure assessment: The birth of a new science. *Journal of Exposure Analysis and Environmental Epidemiology*, **5:** 449–472.

Pamuk, E., Makuc, D., Heck, K., Reuben, C. and Lochner, K. (1998). *Socioeconomic Status and Health Chartbook. Health, United States, 1998.* National Center for Health Statistics, Hyattsville, MD. (http://www.cdc.gov/nchs/data/hus/hus98.pdf)

Paulos, J.A. (1988). *Innumeracy: Mathematical Illiteracy and Its Consequences*. Hill and Wang, New York.

Paulos, J.A. (1995). *A Mathematician Reads the Newspaper*. Basic Books, New York.

Pearson, S.M. and Rose, K.A. (2001). The effect of sampling design on estimating the magnitude and distribution of contaminated sediments in a large reservoir. *Environmetrics,* **12:** 81–102.

Peduzzi, P., Concato, J., Kemper, E., Holford, T.R. and Feinstein, A.R. (1996). A simulation study of the number of events per variable in logistic regression analysis. *Journal of Clinical Epidemiology*, **49:** 1373–1379.

Peduzzi, P., Detre, K. and Gage, A. (1985). Veterans administration cooperative study of medical versus surgical treatment for stable angina—Progress report: Section 2—Design and baseline characteristics. *Progress in Cardiac Disease*, **28:** 235–243.

Pepe, M.S. (2003). *The Statistical Evaluation of Medical Tests for Evaluation and Prediction.* Oxford University Press, Oxford.

Peterson, D.R., van Belle, G. and Chinn, N.M. (1979). Epidemiological comparison of the Sudden Infant Death Syndrome with other major components of infant mortality. *American Journal of Epidemiology*, **110:** 699–707.

Pettitt, A.N. and Siskind, V. (1981). Effect of within–sample dependence on the Mann–Whitney–Wilcoxon statistic. *Biometrika,* **68:** 437–441.

Praetz, P. (1981). A note on the effect of autocorrelation on multiple regression statistics. *Australian Journal of Statistics* **23:** 309–313.

Prentice, R.L. (1988). Correlated binary regression with covariates specific to each binary observation. *Biometrics* **44:** 1033–1048.

Proschan, M.A. and Waclawiw, M.A. (2000). Practical guidelines for multiplicity adjustment in clinical trials. *Controlled Clinical Trials*, **21**: 527–539.

Rivara, F.P., Koepsell, T.D., Grossman, D.C. and Mock, C. (2000). Effectiveness of automatic shoulder belt systems in motor vehice crashes. *Journal of the American Medical Association*, **283**: 2826–2828.

Rothman, K.J. (1986). *Modern Epidemiology*. Little Brown, Boston.

Rothman, K.J. (1990). No adjustments are needed for multiple comparisons. *Epidemiology*, **1**: 43–46.

Rubin, D.B. (1976). Inference and missing data. *Biometrika*, **63**: 581–592.

Rustagi, J.S. and Wolfe, D.A. (editors) (1982). *Teaching of Statistics and Statistical Consulting*. Academic Press, New York.

Sackett, D.L., Richardson, W.S., Rosenberg, W. and Haynes, R.B. (1997). *Evidence-based Medicine: How to Practice and Teach EBM*. Churchill Livingstone, New York.

Salter, L. (1988). *Mandated Science: Science and Scientists in the Making of Standards*. Kluwer Academic Publishers, Dordrecht, Boston and London.

Samet, J.M., Dominici, F., Zeger, S.L., Schwartz, J. and Dockery, D.W. (2000a). *National Morbidity, Mortality, and Air Pollution Study, 2000. Part I: Methods and Methodologic Issues*. Report 94. Health Effects Institute, Boston, MA.

Samet, J.M., Zeger, S.L., Dominici, F., Curriero, F., Coursac, I., Dockery, D.W., Schwartz, J. and Zanobetti, A. (2000b). *National Morbidity, Mortality, and Air Pollution Study, 2000. Part II. National Morbidity, Mortality, and Air Pollution in the United States*. Report 94. Health Effects Institute, Boston, MA.

Sandman, P.M. (1997). Mass media and environmental risk: seven principles. In *What Risk?*, R. Bate (editor). Butterworth-Heinemann, Oxford, U.K., pp.275–284.

Sankoh, A.J., Huque, M.F. and Dubey, S.D. (1997). Some comments on frequently used multiple endpoint adjustment methods in clinical trials. *Statistics in Medicine*, **16**: 2529–2542.

Sankrithi, U., Emanuel, I., and van Belle, G. (1991). Comparison of linear and exponential multivariate models for explaining national infant and child mortality. *International Journal of Epidemiology*, **20**: 565–570.

Sarle, W.W. (1996). Measurement theory: Frequently asked questions. http://www.measurementdevices.com/mttheory.html.

Satten, G.A. and Kupper, L.L. (1990). Sample size determination for pair-matched case-control studies where the goal is interval estimation of the odds ratio. *Journal of Clinical Epidemiology*, **43**: 55–59.

Scheffé, H. (1959). *The Analysis of Variance*. John Wiley and Sons, New York.

Schlesselman, J.J. (1982). *Case-Control Studies*. Oxford University Press, New York, N.Y.

Schoenfeld, D. (1981). The asymptotic properties of nonparametric tests for comparing survival distributions. *Biometrika*, **68**: 316–319.

Schoenfeld, D. (1983). Sample-size formula for the proportional-hazards regression model. *Biometrics*, **39**: 499–503.

Schouten, H.J.A. (2000). Combined evidence from multiple outcomes in a clinical trial. *Journal of Clinical Epidemiology*, **53**: 1137–1144.

Selvin, S. (1996). *Statistical Analysis of Epidemiological Data*. Oxford University Press, New York.

Sherwin, R.P. and Layfield, L.J. (1976). Protein leakage in the lungs of mice exposed to 0.5 pmm nitrogen dioxide: a fluorescence assay for protein. *Archives of Environmental Health*, **31:** 116–118.

Snedecor, G.W. and Cochran, W.G. (1989). *Statistical Methods*, eighth edition. Iowa State University Press, Ames, IA.

Sprott, D.A. and Vogel-Sprott, M.D. (1987). Use of the log odds ratio to assess the reliability of dichotomous questionnaire data. *Applied Psychological Measurement*, **11:** 307–316.

Stevens, S.S. (1946). On the theory of scales of measurement. *Science*, **103:** 677–680. In Mehrens A.M. and Ebel R.L. (editors)(1967). *Principles of Educational and Psychological Measurement*, Rand McNally, Chicago, IL.

Sutherland, P., Rossini, A., Lumley, T., Lewin-Koh, N., Dickerson, J., Cox, Z. and Cook D. (2000). ORCA: A visualization toolkit for high-dimensional data. *Journal of Computational and Graphical Statistics*, **9:** 509–529.

Syrjala, S.E. (2000). Critique on the use of the delta distribution for the analysis of trawl survey data. *ICES Journal of Marine Science*, **57:** 831–842.

Temple, R.J. (1995). A regulatory authority's opinion about surrogate endpoints. In *Clinical Measurement in Drug Evaluation*, Nimmo W.S. and Tucker, G.T. (editors). John Wiley and Sons, New York.

Thomas, D.C. (1981). General relative risk models for survival time and matched case-control analysis. *Biometrics*, **37:** 673–676.

Thorndike, R.M. (1978). *Correlational Procedures for Research*. Gardner Press, New York.

.ufte, E.R. (1983). *The Visual Display of Quantitative Information*. Graphics Press, Cheshire, CT.

Tufte, E.R. (1990). *Envisioning Information*. Graphics Press, Cheshire, CT.

Tufte, E.R. (1997). *Visual Explanations*. Graphics Press, Cheshire, CT.

Vach, W. and Blettner, M. (1998). Missing data in epidemiologic studies. In *Encyclopedia of Biostatistics*, Volume 4, Armitage, P. and Colton, T. (editors). John Wiley and Sons, Chichester, U.K., pp.2637–2654.

van Belle, G. and Arnold, A. (1999). Reliability of cognitive tests used in Alzheimer's disease. *Statistics in Medicine,* **19:** 1411–1420.

van Belle, G. and Friel, P.N. (1986). Problem of spurious correlation in the evaluation of steady-state carbamazapine levels using metabolite data. *Therapeutic Drug Monitoring*, **8:** 177–183.

van Belle, G., Griffith, W.C. and Edland, S.D. (2001). Contributions to composite sampling. *Environmental and Ecological Statistics*, **8:** 171–180.

van Belle, G., Leurgans, S., Friel, P.N., Guo, S. and Yerby, M. (1989). Determination of enzyme or binding constants using generalized linear models with particular reference to Michaelis–Menten models. *Journal of Pharmaceutical Sciences*, **78:** 413–416.

van Belle, G. and Martin, D. (1993). Sample size as a function of coefficient of variation and ratio of means. *The American Statistician*, **47:** 165-167.

van Belle, G., Uhlmann R.F., Hughes, J.P. and Larson, E.B. (1990). Reliability of estimates of change in mental status test performance in senile dementia of the Alzheimer type. *Journal of Clinical Epidemiology*, **43:** 589–595.

Vardeman, S.B. (1992). What about the other intervals? *The American Statistician*, **46:** 193–197.

Velleman, P.F. and Wilkinson, L. (1993). Nominal, ordinal, interval, and ratio typologies are misleading. *The American Statistician*, **47:** 65–72.

Wainer, H. (1997). *Visual Revelations: Graphical Tales of Fate and Deception from Napoleon Bonaparte to Ross Perot*. Springer-Verlag, New York.

Wainer, H. (1998). Rounding tables. *Chance, New Directions for Statistics and Computing,* **11:** 46–50.

Walter, S.D. (2000). Choice of effect measure for epidemiologic data. *Journal of Clincial Epidemiology*, **53:** 931-939.

Weinberg, C.R. (2001). It's time to rehabilitate the *p*-value. *Epidemiology*, **12:** 288–290.

Whitby, K.T. (1978). The physical characteristics of sulfur aerosols. *Atmospheric Environment*, **12:** 135–159.

Wild, C.J. and Seber, G.A.F. (1993). Comparing two proportions from the same survey. *The American Statistician*, **47:** 178–181.

Wilkinson, L. (1999). *The Grammar of Graphics*. Springer, New York.

Wright, N.T. (1999). *The Challenge of Jesus*. Intervarsity Press, Downers Grove, IL.

Yandell, B.S. (1997). *Practical Data Analysis for Designed Experiments*. Chapman and Hall, London.

Young, F.W. (2001). An explanation of the persistent doctor-mortality association. *Journal of Epidemiology and Community Health*, **55:** 80–84.

Zahm, S.H., Fraumeni, J.F., Jr. and Davis, D.L. (1995). The avoidable causes of cancer. In *President's Cancer Panel Conference on Avoidable Causes of Cancer* (April 7–8, 1994). *Environmental Health Perspectives*, **103**(Suppl. 8): 129.

Zeger, S.L. (2000). Analysis of Longitudinal Data. Second Seattle Symposium in Biostatistics: Analysis of Correlated Data. Department of Biostatistics, University of Washington, Seattle, WA.

Zhou, X-H., and Gao, S. (1997). Confidence intervals for the log-normal mean. *Statistics in Medicine*, **16:** 783–790.

Zhou, X-H. and Gao, S. (1998). Estimation of the log-normal mean. *Statistics in Medicine*, **17:** 2251–2264.

Author Index

Adams, M.M., 11, 195
Agresti, A., 53, 56, 195
Ahnn, S., 195
Altman, D.G., 64, 196
American Statistical Association, 186, 195
Anderson, H.R., 201
Anderson, S.J., 195
Anscombe, F.J., 159, 195
Arensberg, D., 198
Armitage, P., 195, 197, 200–201, 204
Arnold, A., 69, 195, 204
Atkinson, A.C., 22, 195
Avison, D., 190
Bailar, J.C., 82, 195
Baker, A., 198
Barcikowski, R.S., 195
Barclay, J., 199
Barker, A.H., 198
Barnett, V., 114, 196–197
Baskerville, R., 190
Bate, R., 123, 196
Bennett, C.A., 110, 196
Benson, K., 2, 196
Bland, J.M., 64, 196
Blettner, M., 148, 204
Boen, J.R., 176, 196
Bowen, W.M., 110, 196
Box, G.E.P., 10, 196
Boyko, E.J., 98, 196
Breslow, N.E., 22, 196

Brunekreef, B., 201
Bryk, A.S., 144, 196
Caccia, D., 197
Chambers, J.M., 153, 196
Chambers, R.L., 5, 196
Chardos, S., 8, 200
Chatfield, C., 176, 196
Chesterton, G.K., 188
Chinn, N.M., 202
Clayton, D., 22, 76, 82, 101, 196
Cleveland, W.S., 153, 196
Cochran, W.G., 8, 32, 196, 204
Cochrane, A.L., 59, 196–197
Cohen, A.J., 201
Cohen, J., 197
Collett, D., 197
Colton, T., 195, 197, 200–201, 204
Concato, J., 2, 197, 202
Conquest, L., 201
Cook D., 204
Cook, R.J., 197, 199
Copas, J.B., 3, 5, 113, 197
Coursac, I., 203
Cox, D.R., 22, 120, 135, 143, 176, 190, 195, 197
Cox, L.H., 197
Cox, Z., 204
Crouch, E.A.C., 123, 197
Cullen, A.C., 114, 117, 125, 197
Curriero, F., 203
David, H.A., 10, 197

Davis, D.L., 205
Davison, A.C., 26, 197
Day, N.E., 196
DeLury, D.B., 130
DeMets, D.L., 198
Detre, K., 202
Dickerson, J., 204
Diggle, P.J., 74, 166, 168, 197
Dockery, D.W., 203
Doll, R., 54, 197
Dominici, F., 203
Dorfman, R., 197
Draper, N., 126, 197
Dubey, S.D., 203
Duran, C.R., 201
Echt, D.S., 93, 198
Edland, S.D., 204
Efron, B., 25–26, 198
Ehrenberg, A.S.C., 157–158, 198
Emanuel, I., 203
Embretson, S.E., 70, 198
EpiTRENDS, 76, 198
Ernster, V., 199
Farewell, V.T., 197, 199
Feinstein, A.R., 202
Fingerhut, L.A., 201
Finney, D.J., 71, 126, 175, 198
Fisher, L.D., 1, 27, 44, 50, 140, 150, 198
Fisher, R.A., 1, 64, 73, 129, 133, 198
Fleiss, J.L., 198
Ford, E.D., 198
Fraser, D.A.S., 108, 198
Fraumeni, J.F., Jr., 205
Free, S.M., Jr., 201
Frey, H.C., 114, 125, 197
Friedman, H.L, 198
Friedman, L.M., 16, 51, 93, 95, 198
Friel, P.N., 204
Furberg, C.D., 198
Gage, A., 202
Gao, S., 106, 205
Gastwirth, J.L., 8, 198
Gehan, E.A., 28, 198
Gelman, A., 72, 198
Gigerenzer, G., 199
Gladman, D.D., 199
Glass, G.V., 8, 198
Grady, D., 199
Griffith, W.C., 204
Grossman, D.C., 203
Gulledge, T.R., Jr., 10, 201
Guo, S., 204
Guttorp, P., 4–5, 90, 103, 114, 197–198
Halfhill, T.R., 132, 198
Hall, J.W., 180, 199
Hand, D.J., 176, 199

Hanley, J.A., 50, 199
Hartz, A.J., 2, 196
Harvey, A.C., 199
Haseman, J.K., 44, 199
Hayes, B., 199
Haynes, R.B., 203
Heck, K., 202
Hertwig, R., 199
Hills, M., 76, 82, 101, 196
Hinkley, D.V., 26, 197
Hoffrage, U., 96, 183, 199
Holder, R.L., 59, 202
Holford, T.R., 202
Hollander, M., 100, 199
Horton, N.J., 148, 199
Horwitz, R.I., 197
Hotelling, H., 199
Hsieh, F.Y., 84, 199
Huff, D., 153, 199
Hughes, J.P., 204
Hunsicker, L.G., 200
Huque, M.F., 203
Hurlbert, S.H., 108, 199
Husted, J.A., 70, 199
IMS Panel, 175, 199
International Statistical Institute, 186, 199
Jackson, G.L., 201
Jenkins, G.M., 196
Johnson, R.A., 199
Kemper, E., 202
Kendall, M.G., 21–22, 90, 199
Kerlikowske, K., 98, 199
Kleinbaum, D.G., 53, 122, 199
Kleiner, B., 196
Kock, N., 190
Koepsell, T.D., 203
Kramarov, E., 164–165, 199
Kronmal, R.A., 59, 200
Kruskal, W.H., 3, 200
Kupper, L.L., 199, 203
Lachin, J.M., 44, 76, 82, 92, 127, 200
Lakatos, E., 33, 200
Lan, S., 200
Land, C.E., 106, 200
Larson, E.B., 204
Lavori, P.W., 195
Layfield, L.J., 139–140, 204
Leibson, P.R., 198
Leisenring, W., 98, 200
Lenth, R., 189
Lentzner, H., 199
Lettenmaier, D.P., 8, 200–201
Leurgans, S., 204
Lewin-Koh, N., 204
Lewis, E.J., 89, 200
Li, H.G., 3, 5, 197

Liang, K-Y., 197, 200
Lin, L.I., 65, 67, 200
Lindsay, S., 199
Lippman-Hand, A., 50, 199
Lipsitz, S.R., 148, 199
Lissitz, R.W., 8, 200
Little, R.J.A., 148, 200
Lochner, K., 202
Loh, W-Y., 64, 200
Looney, S.W., 10, 201
Lord, F.M., 69, 201
Louis, T., 50, 195, 201
Lumley, T., 112, 201, 204
MacKay, A.P., 155, 201
Makuc, D., 202
Mantel, N., 28, 201
Marascuilo, L.A., 100, 201
March, R.H., 3, 201
Marker, D.A., 201
Martin, D., 37, 204
McMichael, A.J., 113, 201
McSweeney, M., 100, 201
Meier, P., 62, 201
Metz, C.E., 99, 201
Millard, S.P., 8, 103, 110–111, 116, 122, 126, 201
Mitchell, L.B., 198
Mock, C., 203
Mode, N.A., 111, 201
Moore, F., 196–197
Mosteller, F., 3, 200
Muller, K.E., 199
Murray, C.J., 111, 201
Murray, G.D., 148, 201
Myers, M., 190
National Research Council, 122, 201
Needham, S.E., 76, 92, 202
Neerchal, N.K., 110–111, 122, 126, 201
Nelder, J.A., 22, 202
Nelson, C.R., 111, 201
Neville, A.M., 59, 202
Neyman, J., 59, 202
Nizam, A., 199
Novick, M.R., 69, 201
O'Hagan, A., 114, 196
Oakes, M., 18, 25, 202
Obias-Manno, D., 198
Ord, J.K., 199
Ott, W.R., 105, 124–125, 202
Pamuk, E., 163, 202
Paulos, J.A., 12–13, 202
Pearson, S.M., 110, 202
Peckham, P.D., 198
Peduzzi, P., 202
Pepe, M.S., 95, 98, 200, 202
Peters, R.W., 198
Peterson, D.R., 78, 202

Pettitt, A.N., 8, 202
Polansky, M., 195
Praetz, P., 8, 202
Prentice, R.L., 94, 202
Proschan, M.A., 150, 202
Raudenbush, S.W., 144, 196
Reid, N., 135, 143, 197
Reise, S.P., 70, 198
Reuben, C., 202
Richardson, W.S., 203
Rivara, F.P., 148, 203
Rohde, R.D., 200
Rooks, R., 199
Rose, K.A., 110, 202
Rosenberg, W., 203
Rossini, A., 204
Rothman, K.J., 18, 101, 203
Rubin, D.B., 148, 200, 203
Rubin, H., 8, 198
Rustagi, J.S., 176, 203
Sackett, D.L., 98, 203
Salter, L., 103, 203
Samet, J.M., 24, 111, 116, 170–171, 203
Sampson, P.D., 197
Sanders, J.R., 198
Sandman, P.M., 123, 203
Sankoh, A.J., 150, 203
Sankrithi, U., 59, 203
Sarle, W.W., 23, 203
Satten, G.A., 203
Saydah, S., 199
Scheffé, H., 135, 203
Scheffé, H., 8, 10
Schlesselman, J.J., 203
Schoenfeld, D., 203
Schouten, H.J.A., 150, 203
Schwartz, J., 203
Seber, G.A.F., 205
Selvin, S., 44, 76, 100, 203
Shah, N., 197
Sheppard, L., 112, 201
Sherwin, R.P., 139–140, 204
Shi, J.Q., 113, 197
Sickles, E.A., 199
Siskind, V., 8, 202
Smith, H., 126, 197
Smith, R.L., 197
Snedecor, G.W., 32, 204
Sprott, D.A., 204
St. Leger, A.S., 196–197
Stevens, J., 195
Stevens, S.S., 23, 204
Stuart, A., 199
Sutherland, P., 170, 204
Syrjala, S.E., 126, 204
Temple, R.J., 93, 204

Thomas, D.C., 204
Thompson, M.L., 114, 197
Thorndike, R.M., 204
Tufte, E.R., 153–155, 158–159, 161–162, 204
Tukey, P.A., 196
Uhlmann R.F., 204
Vach, W., 148, 204
van Belle, G., 1, 27, 37, 44, 50, 69, 110, 140, 150, 198, 202–204
van Belle, L.C., 172
Vardeman, S.B., 121, 205
Velleman, P.F., 23, 205
Vogel-Sprott, M.D., 204
Waclawiw, M.A., 150, 202
Wainer, H., 158, 205
Walter, S.D., 102, 205
Wedderburn, R.W.M., 22, 202
Weeks, J., 199

Weinberg, C.R., 18, 205
Whitby, K.T., 118, 205
Wichern, D.W., 199
Wild, C.J., 205
Wilkinson, L., 5, 23, 205
Wilson, R., 123, 197
Wolfe, D.A., 100, 176, 199, 203
Wood-Harper, T., 190
Wright, N.T., 30, 205
Yandell, B.S., 18, 205
Yearsley, J.R., 201
Yerby, M., 204
Young, F.W., 59, 205
Zahm, S.H., 124, 205
Zahn, D.A., 176, 196
Zanobetti, A., 203
Zeger, S.L., 94, 197, 200, 203, 205
Zhou, X-H., 106, 205

Topic Index

ϕ-coefficient, 63
t-statistic, 8–9, 18
  and sample size, 18
p-value, 16
  asterisk convention, 17
  pitfalls, 17
p-valueS, 6
Acceptable risk, 123
Accuracy, 66
  and precision, 66
Achilles' heel of statistics, 2
Active treatment, 15
Additional observation, 73
Additive factor, 135
Additive model, 100
Additive partitioning, 137
Additive scale, 100
Additivity, 21
Adherence
  in clinical trial, 149
Adverse event, 122
Advisory consulting, 175
Agreement, 53, 64
  components of, 65
  geometric interpretation, 65
AIDS, 25, 93
Air particulate, 107
Air pollution, 115
Air toxics, 104

AIRS
  Aerometric Information Retrieval System, 115
Alternative hypothesis, 29, 32
Alzheimer's disease, 72, 86
Analysis of variance, 15, 57, 108, 138, 140
ANOVA, 140
ANOVA sample size, 190
AR(1) process, 8
ASA, 186
Association, 53
  causal, 55
Assumption of independence, 70
Assumptions
  of classical test theory, 69
Asymptotic relative efficiency, 99
Authority, 12
Autocorrelated data, 9
Automobile air pollution, 113
Background noise, 43
Balance, 136
Balance in assignment of factors, 130
Balanced design, 136
Balanced incomplete block design, 137
Balanced study, 137
Bar graph
  stacked, 164
Baseline observation, 72
Baseline
  adjusting for, 62

for standardization, 62
to increase precision, 61
Bayes' theorem, 183
Bayesian prediction interval, 50
Bernoulli trial, 43
Between-subject variability, 73
Bias, 2, 65
  and accuracy, 68
  and differential scale, 69
  indicated in graph, 65
  location, 65
  regression coefficients, 88
  selection, 112
  systematic, 65
Binary data
  chi-square test for correlation, 62
  differential accuracy, 62
Binary observation, 62
Binary outcome, 29, 75
Binomial distribution, 43
  and sample size, 43
  variance stabilizing transformation, 44
Binomial variable, 23
Bioassay, 70, 126
  design of, 71
Biological variable, 37
Biomarker, 125
  of effect, 125
  of exposure, 125
  of mechanism, 125
Bivariate normal model, 67
Bivariate sampling scheme, 56
Blocking, 131, 134
  agricultural origin, 131
  and multi-center clinical trial, 132
  candidates for, 132
  in clinical trial, 131
  in paired study, 132
Bootstrap, 22, 25–26
Bootstrap estimate
  confidence interval, 106
Bootstrap
  vs. delta method, 126
Box plot, 162
Calibration, 55, 68, 125
  inverse regression, 125
  multivariate, 126
Cannonball, 4
Carry-over effect, 143
Case-control study, 75, 79, 86, 130
  frequency, 82
Categorical data, 56
Categorical variable, 53
Causal chain, 94
Causal covariation, 54
Causal factor, 113

Causation, 54
Cell frequency
  and additivity, 136
Censored observation, 94
Censoring, 91–92
Change
  within individuals, 72
  assessment of, 72
  comparison between two groups, 74
  estimating sample size, 74
  importance of spacing, 72
Chi-square statistic, 7, 23, 136
Classical measurement model, 69
Clinical endpoint
  and surrogate, 92
Clinical outcome
  regulatory interest, 93
  and surrogate, 93
Clinical trial, 15, 129, 150
  active treatment, 15
  and change, 72
  control treatment, 15
  multiple comparison, 149
  surrogates, 93
Clinical trials, 32
Clustering, 77
Coefficient of variation, 35, 106
Cohort study, 79
  and odds ratio, 84
  frequency, 82
  prevalence, 81
Collaborative relationship, 175
Comparative study, 133, 179
Comparing measurements, 66
Comparing risks, 102
Comparison of two binomials
  odds ratio and relative risk, 86
Competing models, 12
Competing risks, 75
Completely randomized design, 138
Compliance, 116
Composite sampling, 110
  double, 111
  relative cost, 111
Computing package
  weak point, 153
Concordance, 53, 67
Conditional independence, 109
Conditional probability, 13
Confidence interval, 16, 120
  for population median, 121
  for correlation, 67
  for median of lognormal, 106
  lognormal mean, 106
  overlap of, 39
Confirmatory analysis, 149

Confirmatory research, 149
Confounding, 112
  seasonal, 112
Constraints
  of study, 143
Constructive critical thinking, 178
Consulting, 176
Consulting session
  structure, 176
Consulting
  active listening, 177
  advice of Sir David Cox, 191
  advisory, 175
  appropriate distinctions, 179
  as collegial activity, 184
  authorship or co-authorship, 184
  choice of statistical package, 185
  code of professional ethics, 186
  communication, 182
  constructive critical thinking, 178
  consultant's use of data, 184
  cross-sectional, 175
  diversity of, 192
  downside to the WEB, 190
  ethics, 186
  focused questions, 178
  handling information, 178
  humility, 178
  illustration of proactive, 187
  indicating commitment, 188
  initial consultation, 178
  interactive, 175
  issue of acknowledgment, 184
  knowledge of investigator, 180
  knowledge of self, 180
  limit of the consultant's knowledge, 180
  longitudinal, 175, 179
  natural units, 183
  pacing, 177
  proactive, 187
  reference text, 185
  referencing the WEB, 190
  role in committee, 188
  seeking advice, 190
  strategies, 182
  tailoring advice, 181
  Type III error, 177
  use of natural units, 182
  use of spreadsheet program, 185
  using the WEB, 189
  valid distinction vs. distinctions valid, 178
  withdrawal from participation, 187
Contagious disease, 77
Contingency table, 23, 136
Continuous measurement, 29
Continuous variable, 53

Control treatment, 15
Corrective action, 116
Correlated error, 8
Correlated observation, 109
Correlated process, 10
Correlation coefficient, 58
Correlation
  invariance of, 64
  and ϕ-coefficient, 63
  and proportion of variance explained, 57
  and reliability
    precision, 69
  and sample size, 60
  and symmetry, 56
  as measure of goodness-of-fit, 63
  canonical, 88
  location and scale, 64
  of ratio variables, 59
  pattern and value, 64
  Pearson product-moment, 25, 56, 119
  serial, 8
  Spearman rank, 119
  spurious, 55
Cost, 47
Counting variable, 23
Covariate, 141
Covariation, 53
  and range of predictor variable, 70
  asymmetric, 55
  cherry picking, 56
  comparison with gold standard, 55
  criteria for selection, 55
  Kappa statistic, 56
  natural measures of, 56
  prediction, 55
Cox regression, 76
Critical value, 31
Cross-sectional consulting, 175
Cross-sectional information, 167
Cross-sectional sampling, 77
Cross-sectional study
  frequency, 82
Crossing, 145
Cut point, 120
Data dimensionality, 166
Data
  skewed, 104
Degrees of freedom, 143
  partitioning of, 143
Delta method, 25, 126
Dependent observation, 10
Descriptive analysis, 3
Design, 132, 134
Design of experiment, 16, 61
Design
  balanced, 136

balanced incomplete block, 137
basic element, 134
completely randomized, 138
crossed, 145
factorial, 133
fractional factorial, 135, 143
nesting, 145
orthogonal, 136
partially balanced incomplete block, 137
single factor study, 133
structure, 142–143
Detection level, 104
Deviance, 65–66
sample, 66
Diagnostics, 5
Dichotomization, 99, 101
and loss of information, 99
and odds ratio, 100
in epidemiology, 100
Difference
and symmetry, 107
Dilution
and lognormal distribution, 105
Disease rate
effect on sample size, 83
Distinction, 178
Distribution of differences, 107
Doll's analysis, 54
Dorfman sampling, 110
Double exponential model, 100
Duration of disease, 75
Ecology, 104
Effect size, 6–7, 38
Effective digits, 157
Effective drug
and Type II error, 150
Efficacy, 150
Endpoint, 29, 149
Environment Canada, 116
Environmental data
detection level, 104
Environmental pollution, 116
Environmental sampling, 110
Environmental studies, 103
Environmental study
and source of variation, 113
importance of lognormal distribution, 104
lognormal model, 104
lognormal process, 105
Environmental variability, 126
Environmental variable, 105
EPA, 104, 116–117
Epidemiologic study type
comparisons, 81
Epidemiology, 104
dichotomization, 100

likelihood approach, 76
Equivalence study, 179
Error in measurement, 116
Error sum of squares, 137
Error
between subjects, 139
ignorable, 126
in normal model, 70
in ratio estimation, 126
within subjects, 139
Errors within subjects, 68
Estimation, 6, 16
Events per variable, 87
Excess risk, 113
Experiment
cost of, 48
Experimental design, 10
Experimental studies, 53
Exploratory analysis, 149–150
Exponential distribution, 89
hazard rate, 90
variance, 89
Exponential
link with Poisson, 89
Exposure, 125
Exposure assessment, 124
and surrogates, 125
quantitative, 124
Exposure status, 77
Exposure
acute, 123, 125
chronic, 123, 125
Extended model, 101
Extrapolation
in risk assessment, 123
Factorial design, 133–134
2×2, 136
limitations, 134
main effect, 134
Factorial structure, 135
Factorial treatment, 142
Factors
relationships between, 134
FDA, 150
objectives, 150
review committee, 150
Finite population correction, 33
First order approximation, 135
Fisher z-transformation, 25, 60, 67
Fixed effect, 21
Fixed effect model, 179
Follow up time
differences in, 92
Food and Drug Administration
see FDA, 150
Four-point assay, 71

Fractional factorial design, 135
Galileo, 3
Generalized Estimating Equation Model (GEE), 22
Generalized Linear Mixed Models (GLMM), 22
Genetic linkage, 15
GIGO, 115
Gold standard
  agreement with, 68
Goodness-of-fit, 63, 67
Goodness-of-fit test, 4
Grant application
  and sample size, 182
Graphical excellence, 159
Graphics
  assessment of non-parallelism, 141
  alternative to bar graph, 162
  alternative to stacked bar graph, 164
  Anscombe regression data, 159
  bar graph, 162
  box plot, 168
  cross-sectional, 168
  cross-sectional data, 167
  factorial design, 141
  high-dimensional data, 170
  interaction plot, 141
  longitudinal, 168
  longitudinal data, 167
  multiple regression, 170
  pie chart, 160
  residual plots, 172
  stacked bar graph, 164
  unnecessary dimensions, 166
Graphing
  amount of ink, 161
  chart junk, 162
  data density, 161
  guidelines for longitudinal data, 168
  immediacy, 161
  informativeness, 162
  linking, 170
  manipulation, 170
  multiple pie charts, 161
  ordering values, 162
  rendering, 170
Graphs
  vs. words and table, 154
Hand-waving, 139
Hazard rate, 90
Hazard ratio, 91
Heterogeneous data, 119
Hierarchical analysis, 141, 144
Hierarchical model, 12, 144
High order interaction, 135
High risk
  dependence on model, 101
Higher order effect, 143

Histogram, 8
Homogeneity of variance, 8, 20–21
  preliminary test, 10
Hypothesis, 14
Hypothesis testing, 16
  appropriate error term, 108
  assumptions, 7
  binary observations, 62
  equipoise, 15
  vs. estimation, 16
Hypothesis
  one-sided, 14
  two-sided, 14
Incidence, 75
  relation to prevalence, 90
Independence
  lack of, 109
Independent random variable, 105
Independent samples, 108
  unequal variance, 46
Ineffective drug
  and Type I error, 150
Inference, 2–3
  nonparametric interval, 121
  validity of, 131
  based on confidence interval, 33
  Bayesian, 24
  confidence interval, 120
  likelihood, 24
  multifactorial cause, 113
  nature of statistical, 25
  Neyman–Pearson, 24
  prediction interval, 120
  problematic, 113
  schools of inference, 24
  statistical, 25
  tolerance interval, 120
Informative model, 120
Initial consultation, 178
Input uncertainty, 114
  and measurement error, 114
Interaction, 134
Interaction effect, 143
Interaction plot, 141
Interaction
  high order, 135
  logarithm of relative risk, 135
  logarithm of the odds ratio, 135
  precision of the estimate of, 134
Interactive consulting, 175
Intercept
  inference interval, 122
Interim analysis, 149
Intermediary response, 125–126
Interpreting statistical output, 179
Interval variable, 23

Intrinsic study, 179
Inverse regression, 125
ISI, 186
Item response theory, 70
  logistic model, 70
Jensen's inequality, 41
Kappa statistic, 7, 56
Latent variable, 94
Law of small numbers, 40
Learning effect, 73
Linear model, 11
  as first approximation, 4
Linear regression
  and association, 70
  and change, 72
Linear trend
  test for, 140
Linearity
  check for, 71
Link, 22
Linkage parameter, 15
Location bias, 65
Location differential, 67
Location shift, 107
Log transformation
  suitability of, 106
Logarithmic transformation
  exponential distribution, 92
Logistic distribution, 75
Logistic model, 88
  item response theory, 70
Logistic regression, 24
  and number of events per variable, 87
  stable estimates, 87
Logit transformation, 97
Lognormal distribution
  as central limit theorem, 105
  for environmental study, 104
  properties, 105
Lognormal mean
  confidence interval, 106
Lognormal process
  environmental study, 105
Longitudinal consulting, 175, 179
  issue of authorship, 184
Longitudinal data
  graphing guidelines, 168
Longitudinal information, 167
Loss of information, 120
Loss to follow-up, 91–92
Lower order effect, 144
Mahalanobis distance, 6
Main effect, 134, 143
Matched control, 62
Matching
  and binary data, 63

assessing effectiveness of, 63
Mean
  median
    coefficient of variation relationship, 106
Meaning of model, 5
Measurement, 114
Measurement error, 94
Measurement scale, 64
Measurement variability, 3
Measurement
  error, 114
  in environmental study, 104
  in time or space, 108
  unit or scale, 117
Mechanism
  action of air pollutant, 113
Median, 108
Median test, 63, 100
  two-sample, 100
Median
  Bernoulli random variable, 108
Metadata, 115
Minimum variance, 46
Missing data, 146–147
  and treatment bias, 148
  complete case analysis, 147
  dental X-ray, 147
  due to treatment side-effects, 148
  how to handle, 146
  reasons for, 146
  sensitivity analysis, 146
Missingness, 147
Mixed effect model, 179
Mixture variables, 23
Model sensitivity, 112
Model uncertainty, 114–115
Model validation, 112
Model
  bivariate normal, 70
  classical test theory, 69
  competing, 12
  components of, 21
  extended, 101
  fixed effect, 179
  for survival, 89
  forecast, 4
  GEE, 22
  hierarchical, 144
  intrinsically nonlinear, 22
  level, 4
  link, 22
  meaning, 5
  mixed effect, 179
  multiplicative, 59
  non-hierarchical, 5
  null, 144

parametric vs. nonparametric, 120
PBPK, 5
physical, 5
random effect, 179
requirement for, 101
simplicity, 11
stepwise procedure, 18
structural, 5
Mortality
  and air pollution, 112
Multiple comparison, 149–150
  in regulatory context, 150
  and drug evaluation, 150
  Bonferroni method, 150
  choice of procedure, 150
  effect on level of significance, 149
  effect on Type I error, 149
  FDA guidelines, 150
  Hochberg procedure, 150
  in observational study, 150
  independent hull hypotheses, 149
  interim analysis, 149
  pairwise comparison, 150
  Scheffé method, 150
  unbalanced data procedure, 150
Multiplication
  of probabilities, 13
Multiplicative model, 100
Multiplicative scale, 100
Multivariate calibration, 126
Multivariate normal data, 56
Natural units, 183
Negative predictive value, 7, 96
Nested design
  example of, 145
Nesting, 145
New drug application, 150
Noise, 21
Non-hierarchical model, 5
Nonexperimental data, 3
Nongovernmental organization, 104
Nonlinearity
  and design, 71
Nonparametric analysis, 119
Nonparametric test, 7
Nonparametric tests
  effect of correlated errors, 8
Nonrandomized study, 130
Normal distribution, 18, 21, 27, 31
Normal model
  and analysis, 51
Normality, 20–21
Normality assumption, 8
  check with histogram, 8
Null hypothesis, 29
  no association, 13

Null model, 144
Number of events, 84
  and precision, 84
Numerical data
  presentation, 154
Observational data, 2
Observational studies, 2, 53
Observational study, 16, 130
Observational unit, 29
Observational vs. experimental studies, 2
Ockham's razor, 11
  various versions, 11
Odds ratio, 7, 56, 76, 78, 102
Odds ratio vs. relative risk, 79
Odds ratio
  in logistic model, 80
  advantages, 79
  and dichotomization, 100
  and relative risk, 81
  definition, 78
  equality with relative risk, 81
  small, 111
Omnibus quantity, 6
  and contingency table, 7
  concordance, 67
  estimation and study design, 6
One-sided test, 15
  ANOVA, 15
  genetic linkage, 15
One-way classification, 137
Ordinal variable, 23, 53
Orthogonal study, 136
Outcome space, 136
Outcome variable, 3
  future value, 122
Outcome
  and surrogate, 93
Outlier, 65, 120, 141, 158
Overlapping confidence intervals, 39
  and test of significance, 39
Pairing data, 61
Pairwise comparison, 150
Paradoxical association, 88
Parameter estimates, 57
Parameter estimation, 6
Parametric analysis, 119
Parametric test, 7
  effect of correlated error, 8
Parsimony, 11, 144
  concept of, 144
Partial surrogate, 94
Particulate, 113, 117
  number vs. area vs. mass, 118
Partitioning
  of degrees of freedom, 143
  additivity, 136

of sum of squares, 136
PCB levels, 114
Pearson product-moment correlation, 25, 119
Person-years, 77
Physical model, 5
Pivotal quantity, 18
Pivotal variable, 18
Plausible mechanism, 113
Poisson distribution, 40, 73
   and sample size, 82
   and time interval, 41
   with noise, 41
Poisson model, 89
   assumptions, 77
   extra-Poisson variation, 77
   for cohort study, 83
   incidence and prevalence, 76–77
Poisson process, 90
Poisson random variable, 21, 41, 108
   sum of, 41
Poisson
   link with exponential, 89
Pollutant
   unit of measurement, 117
Population, 1, 3
Population mean, 6, 32
Population modeling
   data selection process, 5
Population of interest, 109
Population
   1800 to 2000, 168
   representative sample, 109
   US by region and year, 168
Positive predictive value, 7, 96
Precision, 66
   and unequal variance, 46
Preclinical screening, 75
Prediction, 4
Prediction interval, 120
Predictive value
   in screening, 95
Predictor variable, 3, 55, 70, 122
   and proportion of variance explained, 58
   calibration, 126
Prevalence, 7, 75, 95
Prevalence study, 77
Prevalence
   and positive predictive value, 96
   domination in screening, 97
   relation to incidence, 90
Primary endpoint, 149
Principal component analysis, 88
Proactive statistical consulting, 187
Probability, 2, 13
Probability of dropout, 73
Probability

   meaning of, 13
   and multiplication, 13
   conditional, 13, 79
   court case example, 12, 14
   relative frequency interpretation, 24
   subjective interpretation, 24
Proportion of variance explained, 57–58
Proportion testing, 63
Proportion
   and median test, 63
Prospective study, 77
Pseudoreplication, 108, 114
Publication bias, 113
Quadratic terms, 11
R.A. Fisher
   and correlation, 64
Radon, 123
Random effect, 21
Random effect model, 179
Randomization, 10, 129, 131, 134
   and analysis, 137
   and analysis of variance, 130
   and confidence interval, 130
   and test of significance, 130
   result of lack of, 130
   three-fold purpose, 130
Randomized block, 131, 138
Randomized clinical trial, 2
Randomness, 2
Rank statistic, 53
Ranked set sampling, 110
Rare disease assumption, 75
Rare events, 40
Rate ratio, 100
Rates
   vs. differences, 100
Ratio variable, 23
Ratio
   hazard, 91
Receiver Operating Characteristic Curve, 98
Reductionism, 12
Regression, 53
Regression model, 120, 122, 126
Regression sampling scheme, 56–57
Regression
   and ANOVA, 57
   residuals, 71
   and asymmetry, 56
   and design, 58
   Cox, 76
   future observation, 122
Rejection region, 15, 31
Relative risk, 7, 76
   and odds ratio, 81
   definition, 78
   equality with odds ratio, 81

property of, 80
Reliability
  and precision, 69
Repeated observation, 73
Replication, 108, 143
  inadequate, 109
Representativeness
  of the population, 109
  qualitative criterion, 109
Reproducibility
  of a test, 68
Rescaling, 65
Research question, 115
Responsiveness
  of test, 70
Retrospective study, 77
Richter scale, 20
Risk, 122
Risk assessment, 104, 114, 122
  eating shell fish, 14
  accountability, 124
  dose-response, 122
  exposure assessment, 122
  hazard identification, 122
  radon, 123
  risk characterization, 122
  risk management, 122
  toxicology, 123
Risk difference, 102
Risk factor
  and coronary disease, 95
Risk factors, 75
  and additive models, 101
Risk ratio, 102
Risk
  as adverse event, 122
  as probability, 122
ROC curve, 98
Rule of threes
  illustration, 49
  in binomial case, 50
  Poisson basis of, 49
Sample correlation
  and slope, 67
Sample mean, 6, 34
Sample size, 33
Sample size estimation, 33
Sample size formula, 35, 37
Sample size
  and correlation, 60
  and logarithm of the odds ratio, 84
  in cohort studies, 85
  adjusting for dropouts, 92
  and analysis, 50
  and binomial distribution, 43
  and coefficient of variation, 36

  and hazard rate, 91
  and number of events, 82
  and percentage change, 36
  and precision, 18
  and prevalence, 83
  and the number of variables, 87–88
  based on odds ratio, 85
  for cohort study, 83
  generic, 182
  square root rule, 157
  unequal, 45
  what can go wrong, 88
  with cost consideration, 47
Sample space, 13
Sample
  representative, 3
Sampling, 1
Sampling design, 110
Sampling scheme, 56
  bivariate, 56
  regression, 56
Sampling strategy, 139
Sampling without replacement, 33
Sampling
  composite, 110
  convenience sample, 2
  cross-sectional, 77
  ranked set, 110
  regulatory requirements for, 109
  simple random, 109
  survey, 2
SAS statistical package, 136
Scale differential, 67
Scale of measurement, 23
Scale variability, 69
Scale
  differential, 65
  of measurement, 20
Scatterplot
  to assess agreement, 65
  to detect bias, 65
Science
  and values, 104
Screening effectiveness, 183
Screening
  domination of prevalence, 97
Secondary analysis, 115
Secondary endpoint, 149
Secondary standard, 117
Selection bias, 5, 112
Sensitivity, 7, 95
Sensitivity analysis, 146
Sensitivity to assumptions, 112
  small effects in large populations, 112
Sentence table, 154
Sequential measurement, 8

Serial correlation, 8, 10
SIDS, 78
Sign test, 99
  asymptotic relative efficiency, 99
Signal, 21
Significant digits, 157
  deep background, 158
  intermediary output, 158
Simple random sampling, 109
Single factor design, 133
Size of test, 30
Skewed data
  lognormal model, 104
Skewness, 107
Slope comparisons, 67
Slope measurement, 64
Small data set, 26
Small effects, 112
Sole statistician, 181
Sound science, 103
Source of variation, 113, 126
Spatial heterogeneity, 110
Spearman rank correlation, 119
Specificity, 7, 95
Spurious correlation, 55, 58
  stork example, 59
Square root transformation, 83, 108
  exponential variable, 90
Staged sampling, 110
Standard, 117
Standard deviation, 27
Standard error, 26, 34
  and range of predictor variable, 70
  estimation in small data set, 26
  of slope, 70
Standard
  and measurement, 116
Standardization of measurement, 125
Standardized difference, 33
STAR program, 103
Statement of risk
  sample space, 13
Statistical consultant, 1, 29
Statistical consulting *see* Consulting, 175
Statistical diagnostics, 5
Statistical immaturity, 25
Statistical independence, 21
Statistical power, 10
Statistical power, 30, 32, 51
  and efficiency, 99
  for distinguishing models, 101
  in cohort studies, 85
Statistics
  as the methodology of the sciences, 192
Stepwise regression, 17
Stochastic process, 90

Stochastic variable, 116
Stratified sampling, 110
Structural model, 5
Study constraint, 143
Study unit, 143
Study
  analysis, 150
  case-control, 79, 86
  cohort, 79
  comparative, 179
  conduct, 150
  cross-sectional, 77
  design, 150
  equivalence, 179
  intrinsic, 179
  presentation, 150
  prospective, 77
  purpose, 150
  retrospective, 77
Subpopulation
  sensitive, 117
Summary statistic, 7
Surrogate, 93
  partial, 94
  regulatory issue, 95
Survey sample, 33
Survey sampling, 110
Survival analysis, 75, 94
Survival patterns, 75
Survival
  modeled by exponential distribution, 89
Symmetric vs. asymmetric, 54
Symmetry transformation, 107
Systematic effect, 21
  reduced by blocking, 131
Systematic sampling, 110
Table
  almanac, 157
  biological measurements, 157
  census data, 157
  coefficient of variation, 157
  construction, 158
  continuous measurements, 157
  effective digits, 157
  number of significant digits, 157
  precision of percentages, 157
  reason for use, 154
  structure, 155
  vs. words and graph, 154
  white space, 156
Taxonomy of measures, 55
Test–retest model, 69
Test of significance
  one-sided, 15
Test reliability, 68
Test

one-sided, 15
Third moment about the mean, 107
Three–number summary, 7
Three-point assay, 71
Tolerance interval, 120
Toxicology, 104, 123
Transformation, 21, 135
  of response, 135
Treatment effect, 29, 35, 94, 143
  and scale of measurement, 51
  relative, 35
  size of, 29
Treatment sum of squares, 137
Treatment vs. design structure, 142
Treatment
  balanced allocation, 136
  structure, 142–143
TSP, 118
Two-sample *t*-test
  efficiency, 99
Type I error, 7, 29, 32, 149
  affected by correlated error, 8
  and interim analysis, 150
  effect of autocorrelation, 9
Type II error, 29, 32
  and statistical power, 30
Type III error, 177
Uncertainty, 114
  input, 114
  model, 114
Unequal sample size, 45
Unequal variance
  and sample size, 46
Unethical practice, 15
Uniform distribution, 27, 110
Uniform prior, 50
Unit of measurement, 117
Unit
  science of, 20
Units, 19
  natural, 19
Unnecessary hypotheses, 11
Valid inference, 109
Validity, 131

Validity of result, 130
Validity
  and accuracy, 69
  of test, 69
Variability, 29, 32–33, 35, 37, 113
  between-subject, 73
  environmental, 126
  estimate of, 7
  in response, 132
  partitioning, 21
  systematic sources of, 129
  vs. uncertainty, 114
  within-treatment, 131
Variable transformation, 21
  for Poisson variable, 22
  homogeneity of variance, 22
  normalization, 21
  questions to ask, 22
  rank order of arithmetic mean vs. geometric
    mean, 22
  side effects, 21
  symmetrization, 21
  variance stabilization, 21
Variable
  type, 23
Variance homogeneity, 31
Variance stabilizing transformation, 44, 83, 108
  exponential distribution, 90
Variance
  of the difference, 61
  control of, 71
  delta method, 126
  homogeneity of, 8
  large sample estimates, 88
  number of controls per case, 86
  of differences, 133
  proportion explained, 57
  within subject, 68
Variation, 1
Wald statistic, 88
Words
  vs. table and graph, 154
Youden's *J* statistic, 7
Yule's *Q* statistic, 7

BERRY, CHALONER, and GEWEKE · Bayesian Analysis in Statistics and Econometrics: Essays in Honor of Arnold Zellner

BERNARDO and SMITH · Bayesian Theory

BHAT · Elements of Applied Stochastic Processes, *Second Edition*

BHATTACHARYA and JOHNSON · Statistical Concepts and Methods

BHATTACHARYA and WAYMIRE · Stochastic Processes with Applications

BILLINGSLEY · Convergence of Probability Measures, *Second Edition*

BILLINGSLEY · Probability and Measure, *Third Edition*

BIRKES and DODGE · Alternative Methods of Regression

BLISCHKE AND MURTHY · Reliability: Modeling, Prediction, and Optimization

BLOOMFIELD · Fourier Analysis of Time Series: An Introduction, *Second Edition*

BOLLEN · Structural Equations with Latent Variables

BOROVKOV · Ergodicity and Stability of Stochastic Processes

BOULEAU · Numerical Methods for Stochastic Processes

· Bayesian Inference in Statistical Analysis

· R. A. Fisher, the Life of a Scientist

BOX and DRAPER · Empirical Model-Building and Response Surfaces

*BOX and DRAPER · Evolutionary Operation: A Statistical Method for Process Improvement

BOX, HUNTER, and HUNTER · Statistics for Experimenters: An Introduction to Design, Data Analysis, and Model Building

BOX and LUCEÑO · Statistical Control by Monitoring and Feedback Adjustment

BRANDIMARTE · Numerical Methods in Finance: A MATLAB-Based Introduction

BROWN and HOLLANDER · Statistics: A Biomedical Introduction

BRUNNER, DOMHOF, and LANGER · Nonparametric Analysis of Longitudinal Data in Factorial Experiments

BUCKLEW · Large Deviation Techniques in Decision, Simulation, and Estimation

CAIROLI and DALANG · Sequential Stochastic Optimization

CHATTERJEE and HADI · Sensitivity Analysis in Linear Regression

CHATTERJEE and PRICE · Regression Analysis by Example, *Third Edition*

CHERNICK · Bootstrap Methods: A Practitioner's Guide

CHILÈS and DELFINER · Geostatistics: Modeling Spatial Uncertainty

CHOW and LIU · Design and Analysis of Clinical Trials: Concepts and Methodologies

CLARKE and DISNEY · Probability and Random Processes: A First Course with Applications, *Second Edition*

COCHRAN and COX · Experimental Designs, *Second Edition*

CONGDON · Bayesian Statistical Modelling

CONOVER · Practical Nonparametric Statistics, *Second Edition*

COOK · Regression Graphics

COOK and WEISBERG · Applied Regression Including Computing and Graphics

COOK and WEISBERG · An Introduction to Regression Graphics

CORNELL · Experiments with Mixtures, Designs, Models, and the Analysis of Mixture Data, *Third Edition*

COVER and THOMAS · Elements of Information Theory

· A Handbook of Introductory Statistical Methods

· Planning of Experiments

CRESSIE · Statistics for Spatial Data, *Revised Edition*

CSÖRGÖ and HORVÁTH · Limit Theorems in Change Point Analysis

DANIEL · Applications of Statistics to Industrial Experimentation

DANIEL · Biostatistics: A Foundation for Analysis in the Health Sciences, *Sixth Edition*

*DANIEL · Fitting Equations to Data: Computer Analysis of Multifactor Data, *Second Edition*

DAVID · Order Statistics, *Second Edition*

*Now available in a lower priced paperback edition in the Wiley Classics Library.

*DEGROOT, FIENBERG, and KADANE · Statistics and the Law
DETTE and STUDDEN · The Theory of Canonical Moments with Applications in Statistics, Probability, and Analysis
DEY and MUKERJEE · Fractional Factorial Plans
DILLON and GOLDSTEIN · Multivariate Analysis: Methods and Applications
DODGE · Alternative Methods of Regression
*DODGE and ROMIG · Sampling Inspection Tables, *Second Edition*
*DOOB · Stochastic Processes
DOWDY and WEARDEN · Statistics for Research, *Second Edition*
DRAPER and SMITH · Applied Regression Analysis, *Third Edition*
DRYDEN and MARDIA · Statistical Shape Analysis
DUDEWICZ and MISHRA · Modern Mathematical Statistics
DUNN and CLARK · Applied Statistics: Analysis of Variance and Regression, *Second Edition*
DUNN and CLARK · Basic Statistics: A Primer for the Biomedical Sciences, *Third Edition*
DUPUIS and ELLIS · A Weak Convergence Approach to the Theory of Large Deviations
*ELANDT-JOHNSON and JOHNSON · Survival Models and Data Analysis
ETHIER and KURTZ · Markov Processes: Characterization and Convergence
EVANS, HASTINGS, and PEACOCK · Statistical Distributions, *Third Edition*
FELLER · An Introduction to Probability Theory and Its Applications, Volume I, *Third Edition,* Revised; Volume II, *Second Edition*
FISHER and VAN BELLE · Biostatistics: A Methodology for the Health Sciences
*FLEISS · The Design and Analysis of Clinical Experiments
FLEISS · Statistical Methods for Rates and Proportions, *Second Edition*
FLEMING and HARRINGTON · Counting Processes and Survival Analysis
FULLER · Introduction to Statistical Time Series, *Second Edition*
FULLER · Measurement Error Models
GALLANT · Nonlinear Statistical Models
GHOSH, MUKHOPADHYAY, and SEN · Sequential Estimation
GIFI · Nonlinear Multivariate Analysis
GLASSERMAN and YAO · Monotone Structure in Discrete-Event Systems
GNANADESIKAN · Methods for Statistical Data Analysis of Multivariate Observations, *Second Edition*
GOLDSTEIN and LEWIS · Assessment: Problems, Development, and Statistical Issues
GREENWOOD and NIKULIN · A Guide to Chi-Squared Testing
GROSS and HARRIS · Fundamentals of Queueing Theory, *Third Edition*
*HAHN · Statistical Models in Engineering
HAHN and MEEKER · Statistical Intervals: A Guide for Practitioners
HALD · A History of Probability and Statistics and their Applications Before 1750
HALD · A History of Mathematical Statistics from 1750 to 1930
HAMPEL · Robust Statistics: The Approach Based on Influence Functions
HANNAN and DEISTLER · The Statistical Theory of Linear Systems
HEIBERGER · Computation for the Analysis of Designed Experiments
HEDAYAT and SINHA · Design and Inference in Finite Population Sampling
HELLER · MACSYMA for Statisticians
HINKELMAN and KEMPTHORNE: · Design and Analysis of Experiments, Volume 1: Introduction to Experimental Design
HOAGLIN, MOSTELLER, and TUKEY · Exploratory Approach to Analysis of Variance
HOAGLIN, MOSTELLER, and TUKEY · Exploring Data Tables, Trends and Shapes
*HOAGLIN, MOSTELLER, and TUKEY · Understanding Robust and Exploratory Data Analysis
HOCHBERG and TAMHANE · Multiple Comparison Procedures

*Now available in a lower priced paperback edition in the Wiley Classics Library.

HOCKING · Methods and Applications of Linear Models: Regression and the Analysis of Variables

HOEL · Introduction to Mathematical Statistics, *Fifth Edition*

HOGG and KLUGMAN · Loss Distributions

HOLLANDER and WOLFE · Nonparametric Statistical Methods, *Second Edition*

HOSMER and LEMESHOW · Applied Logistic Regression, *Second Edition*

HOSMER and LEMESHOW · Applied Survival Analysis: Regression Modeling of Time to Event Data

HØYLAND and RAUSAND · System Reliability Theory: Models and Statistical Methods

HUBER · Robust Statistics

HUBERTY · Applied Discriminant Analysis

HUNT and KENNEDY · Financial Derivatives in Theory and Practice

HUSKOVA, BERAN, and DUPAC · Collected Works of Jaroslav Hajek— with Commentary

IMAN and CONOVER · A Modern Approach to Statistics

JACKSON · A User's Guide to Principle Components

JOHN · Statistical Methods in Engineering and Quality Assurance

JOHNSON · Multivariate Statistical Simulation

JOHNSON and BALAKRISHNAN · Advances in the Theory and Practice of Statistics: A Volume in Honor of Samuel Kotz

JUDGE, GRIFFITHS, HILL, LÜTKEPOHL, and LEE · The Theory and Practice of Econometrics, *Second Edition*

JOHNSON and KOTZ · Distributions in Statistics

JOHNSON and KOTZ (editors) · Leading Personalities in Statistical Sciences: From the Seventeenth Century to the Present

JOHNSON, KOTZ, and BALAKRISHNAN · Continuous Univariate Distributions, Volume 1, *Second Edition*

JOHNSON, KOTZ, and BALAKRISHNAN · Continuous Univariate Distributions, Volume 2, *Second Edition*

JOHNSON, KOTZ, and BALAKRISHNAN · Discrete Multivariate Distributions

JOHNSON, KOTZ, and KEMP · Univariate Discrete Distributions, *Second Edition*

JUREČKOVÁ and SEN · Robust Statistical Procedures: Aymptotics and Interrelations

JUREK and MASON · Operator-Limit Distributions in Probability Theory

KADANE · Bayesian Methods and Ethics in a Clinical Trial Design

KADANE AND SCHUM · A Probabilistic Analysis of the Sacco and Vanzetti Evidence

KALBFLEISCH and PRENTICE · The Statistical Analysis of Failure Time Data

KASS and VOS · Geometrical Foundations of Asymptotic Inference

KAUFMAN and ROUSSEEUW · Finding Groups in Data: An Introduction to Cluster Analysis

KENDALL, BARDEN, CARNE, and LE · Shape and Shape Theory

KHURI · Advanced Calculus with Applications in Statistics

KHURI, MATHEW, and SINHA · Statistical Tests for Mixed Linear Models

KLUGMAN, PANJER, and WILLMOT · Loss Models: From Data to Decisions

KLUGMAN, PANJER, and WILLMOT · Solutions Manual to Accompany Loss Models: From Data to Decisions

KOTZ, BALAKRISHNAN, and JOHNSON · Continuous Multivariate Distributions, Volume 1, *Second Edition*

KOTZ and JOHNSON (editors) · Encyclopedia of Statistical Sciences: Volumes 1 to 9 with Index

KOTZ and JOHNSON (editors) · Encyclopedia of Statistical Sciences: Supplement Volume

KOTZ, READ, and BANKS (editors) · Encyclopedia of Statistical Sciences: Update Volume 1

KOTZ, READ, and BANKS (editors) · Encyclopedia of Statistical Sciences: Update Volume 2

*Now available in a lower priced paperback edition in the Wiley Classics Library.

KOVALENKO, KUZNETZOV, and PEGG · Mathematical Theory of Reliability of Time-Dependent Systems with Practical Applications

LACHIN · Biostatistical Methods: The Assessment of Relative Risks

LAD · Operational Subjective Statistical Methods: A Mathematical, Philosophical, and Historical Introduction

LAMPERTI · Probability: A Survey of the Mathematical Theory, *Second Edition*

LANGE, RYAN, BILLARD, BRILLINGER, CONQUEST, and GREENHOUSE · Case Studies in Biometry

LARSON · Introduction to Probability Theory and Statistical Inference, *Third Edition*

LAWLESS · Statistical Models and Methods for Lifetime Data

LAWSON · Statistical Methods in Spatial Epidemiology

LE · Applied Categorical Data Analysis

LE · Applied Survival Analysis

LEE · Statistical Methods for Survival Data Analysis, *Second Edition*

LePAGE and BILLARD · Exploring the Limits of Bootstrap

LEYLAND and GOLDSTEIN (editors) · Multilevel Modelling of Health Statistics

LINDVALL · Lectures on the Coupling Method

LINHART and ZUCCHINI · Model Selection

LITTLE and RUBIN · Statistical Analysis with Missing Data

LLOYD · The Statistical Analysis of Categorical Data

MAGNUS and NEUDECKER · Matrix Differential Calculus with Applications in Statistics and Econometrics, *Revised Edition*

MALLER and ZHOU · Survival Analysis with Long Term Survivors

MALLOWS · Design, Data, and Analysis by Some Friends of Cuthbert Daniel

MANN, SCHAFER, and SINGPURWALLA · Methods for Statistical Analysis of Reliability and Life Data

MANTON, WOODBURY, and TOLLEY · Statistical Applications Using Fuzzy Sets

MARDIA and JUPP · Directional Statistics

MASON, GUNST, and HESS · Statistical Design and Analysis of Experiments with Applications to Engineering and Science

McCULLOCH and SEARLE · Generalized, Linear, and Mixed Models

McFADDEN · Management of Data in Clinical Trials

McLACHLAN · Discriminant Analysis and Statistical Pattern Recognition

McLACHLAN and KRISHNAN · The EM Algorithm and Extensions

McLACHLAN and PEEL · Finite Mixture Models

McNEIL · Epidemiological Research Methods

MEEKER and ESCOBAR · Statistical Methods for Reliability Data

MEERSCHAERT and SCHEFFLER · Limit Distributions for Sums of Independent Random Vectors: Heavy Tails in Theory and Practice

*MILLER · Survival Analysis, *Second Edition*

MONTGOMERY, PECK, and VINING · Introduction to Linear Regression Analysis, *Third Edition*

MORGENTHALER and TUKEY · Configural Polysampling: A Route to Practical Robustness

MUIRHEAD · Aspects of Multivariate Statistical Theory

MURRAY · X-STAT 2.0 Statistical Experimentation, Design Data Analysis, and Nonlinear Optimization

MYERS and MONTGOMERY · Response Surface Methodology: Process and Product Optimization Using Designed Experiments, *Second Edition*

MYERS, MONTGOMERY, and VINING · Generalized Linear Models. With Applications in Engineering and the Sciences

NELSON · Accelerated Testing, Statistical Models, Test Plans, and Data Analyses

NELSON · Applied Life Data Analysis

NEWMAN · Biostatistical Methods in Epidemiology

*Now available in a lower priced paperback edition in the Wiley Classics Library.

OCHI · Applied Probability and Stochastic Processes in Engineering and Physical Sciences

OKABE, BOOTS, SUGIHARA, and CHIU · Spatial Tesselations: Concepts and Applications of Voronoi Diagrams, *Second Edition*

OLIVER and SMITH · Influence Diagrams, Belief Nets and Decision Analysis

PANKRATZ · Forecasting with Dynamic Regression Models

PANKRATZ · Forecasting with Univariate Box-Jenkins Models: Concepts and Cases

*PARZEN · Modern Probability Theory and Its Applications

PEÑA, TIAO, and TSAY · A Course in Time Series Analysis

PIANTADOSI · Clinical Trials: A Methodologic Perspective

PORT · Theoretical Probability for Applications

POURAHMADI · Foundations of Time Series Analysis and Prediction Theory

PRESS · Bayesian Statistics: Principles, Models, and Applications

PRESS and TANUR · The Subjectivity of Scientists and the Bayesian Approach

PUKELSHEIM · Optimal Experimental Design

PURI, VILAPLANA, and WERTZ · New Perspectives in Theoretical and Applied Statistics

PUTERMAN · Markov Decision Processes: Discrete Stochastic Dynamic Programming

· Linear Statistical Inference and Its Applications, *Second Edition*

RENCHER · Linear Models in Statistics

RENCHER · Methods of Multivariate Analysis, *Second Edition*

RENCHER · Multivariate Statistical Inference with Applications

RIPLEY · Spatial Statistics

RIPLEY · Stochastic Simulation

ROBINSON · Practical Strategies for Experimenting

ROHATGI and SALEH · An Introduction to Probability and Statistics, *Second Edition*

ROLSKI, SCHMIDLI, SCHMIDT, and TEUGELS · Stochastic Processes for Insurance and Finance

ROSS · Introduction to Probability and Statistics for Engineers and Scientists

ROUSSEEUW and LEROY · Robust Regression and Outlier Detection

RUBIN · Multiple Imputation for Nonresponse in Surveys

RUBINSTEIN · Simulation and the Monte Carlo Method

RUBINSTEIN and MELAMED · Modern Simulation and Modeling

RYAN · Modern Regression Methods

RYAN · Statistical Methods for Quality Improvement, *Second Edition*

SALTELLI, CHAN, and SCOTT (editors) · Sensitivity Analysis

*SCHEFFE · The Analysis of Variance

SCHIMEK · Smoothing and Regression: Approaches, Computation, and Application

SCHOTT · Matrix Analysis for Statistics

SCHUSS · Theory and Applications of Stochastic Differential Equations

SCOTT · Multivariate Density Estimation: Theory, Practice, and Visualization

*SEARLE · Linear Models

SEARLE · Linear Models for Unbalanced Data

SEARLE · Matrix Algebra Useful for Statistics

SEARLE, CASELLA, and McCULLOCH · Variance Components

SEARLE and WILLETT · Matrix Algebra for Applied Economics

SEBER · Linear Regression Analysis

SEBER · Multivariate Observations

SEBER and WILD · Nonlinear Regression

SENNOTT · Stochastic Dynamic Programming and the Control of Queueing Systems

*SERFLING · Approximation Theorems of Mathematical Statistics

SHAFER and VOVK · Probability and Finance: It's Only a Game!

SMALL and McLEISH · Hilbert Space Methods in Probability and Statistical Inference

STAPLETON · Linear Statistical Models

STAUDTE and SHEATHER · Robust Estimation and Testing

*Now available in a lower priced paperback edition in the Wiley Classics Library.

STOYAN, KENDALL, and MECKE · Stochastic Geometry and Its Applications, *Second Edition*

STOYAN and STOYAN · Fractals, Random Shapes and Point Fields: Methods of Geometrical Statistics

STYAN · The Collected Papers of T. W. Anderson: 1943–1985

SUTTON, ABRAMS, JONES, SHELDON, and SONG · Methods for Meta-Analysis in Medical Research

TANAKA · Time Series Analysis: Nonstationary and Noninvertible Distribution Theory

THOMPSON · Empirical Model Building

THOMPSON · Sampling

THOMPSON · Simulation: A Modeler's Approach

THOMPSON and SEBER · Adaptive Sampling

TIAO, BISGAARD, HILL, PEÑA, and STIGLER (editors) · Box on Quality and Discovery: with Design, Control, and Robustness

TIERNEY · LISP-STAT: An Object-Oriented Environment for Statistical Computing and Dynamic Graphics

TSAY · Analysis of Financial Time Series

UPTON and FINGLETON · Spatial Data Analysis by Example, Volume II: Categorical and Directional Data

VAN BELLE · Statistical Rules of Thumb

VIDAKOVIC · Statistical Modeling by Wavelets

WEISBERG · Applied Linear Regression, *Second Edition*

WELSH · Aspects of Statistical Inference

WESTFALL and YOUNG · Resampling-Based Multiple Testing: Examples and Methods for p-Value Adjustment

WHITTAKER · Graphical Models in Applied Multivariate Statistics

WINKER · Optimization Heuristics in Economics: Applications of Threshold Accepting

WONNACOTT and WONNACOTT · Econometrics, *Second Edition*

WOODING · Planning Pharmaceutical Clinical Trials: Basic Statistical Principles

WOOLSON and CLARKE · Statistical Methods for the Analysis of Biomedical Data, *Second Edition*

WU and HAMADA · Experiments: Planning, Analysis, and Parameter Design Optimization

YANG · The Construction Theory of Denumerable Markov Processes

*ZELLNER · An Introduction to Bayesian Inference in Econometrics

ZHOU, OBUCHOWSKI, and McCLISH · Statistical Methods in Diagnostic Medicine